Contents

Photocopy Masters

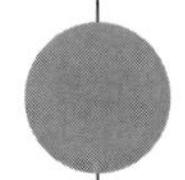

Introduction

The Abacus model

The Abacus materials are designed and written to allow for a daily, structured mathematics lesson:

Mental Warm-up Activities

- Rehearsing previously-taught strategies
- Counting
- Number facts

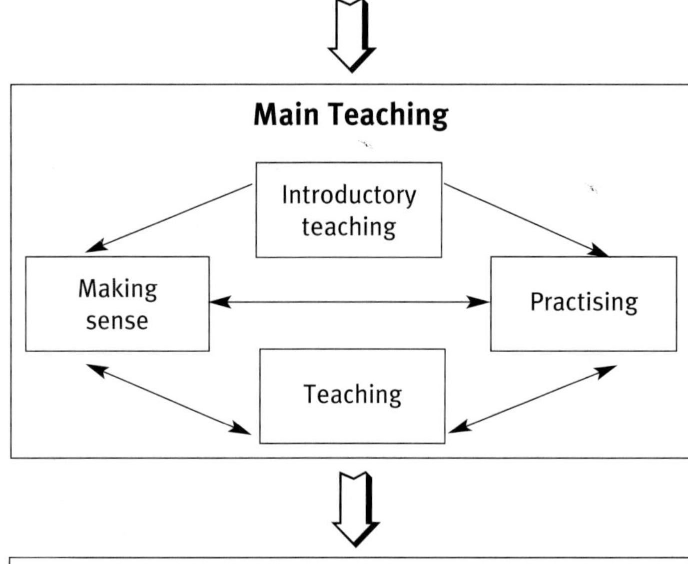

Plenaries

- Reinforcing key skills
- Addressing common difficulties
- Concluding with feedback or an activity

Each day's teaching begins with a whole class mental maths activity. These are presented in the **Mental Warm-up Activities** book.

The main part of the lesson is supported by the **Teacher Cards**:

- the front of each card gives support for whole class teaching for the first day of a topic
- the back of each card gives advice on further teaching for subsequent days and references to the practical activities included in this book. References to relevant **Textbook** pages and **Photocopy Masters** are also provided.

Each lesson is rounded-off with a plenary session. Guidance on key points to reiterate, common misconceptions to look out for, and whole class activities are included on the back of each Teacher Card.

Activity Book

The activities in this book are intended as follow-up to the introductory teaching. Each unit of the programme is supported by a range of activities covering different styles, numbers of children, resources etc.

Each activity includes the following information:

- Appropriate number of children, e.g. pairs, 3–4 children, whole class.
- A list of relevant materials. Any 'specialist' resources, e.g. number grids, number tracks … are provided as Photocopy Masters at the back of the book. Number lines, number cards, place-value cards etc are also available separately in the **Resource Bank**.
- Level of difficulty, indicated by the following codes:

  basic work   for all children   enrichment and extension.
- Learning points are also provided, drawing on the teaching objectives from the Teacher Card. These learning points will assist the teacher in directing the group and making informal assessments. They are also useful as key points to highlight in the plenary session – it may be beneficial to give the children some points to consider when setting up the activity. This will give them a clear focus for the outcome of the activity and any key points they might raise in the plenary session.
- A number of ICT activities are included (these assume some prior knowledge of relevant software, e.g. spreadsheets).

Following the initial teaching input, it is often a useful strategy to carry out some immediate consolidation, with the whole class or large group, working on an appropriate activity. Many of the units include a whole class activity specifically written to support this strategy. Such activities are indicated by the icon:

We suggest that following the initial teaching input for the whole class activity, the children are arranged in pairs or small groups, and work independently for a short period exploring or consolidating their learning. These activities will sometimes lead into the plenary session, where the topic can be rounded-off with a discussion about what the children have learned and any difficulties they encountered.

Group activites

Activities labelled 'Group Activity' use the photocopy master provided on the facing page, along with any materials listed. The instructions for these photocopiable activities, which are provided to supplement those communicated through the teacher, are described in language that is easily interpreted by the children. This enables the activities to be presented directly to the pupils, allowing you to focus your attention on other groups.

Classroom management

When working with groups it is important to have a manageable number of groups (about four is ideal). It may be appropriate for one or two of those groups to be working from a Textbook or Photocopy Master page. You may decide to have more than one group working on the same activity concurrently. You should try to focus your attention on one or two groups, working intensively with them, directing, discussing, evaluating etc. The Abacus model assists management by ensuring that all the children are broadly working within the same topic, at the same time as providing differentiated work through the activities in this book and the Photocopy Masters.

The activities are written with enough detail covering resources and learning outcomes (as well as the description of the activities themselves) to allow any support staff to manage easily groups you are not working with directly.

N1 Place-value

ACTIVITY 1
Whole class, in pairs

- *Rounding a 4-digit number to its nearest ten, hundred and thousand*

Place-value cards (Th, H, T, U) (PCMs 1 to 4)

Shuffle each set of cards and place them face down in four piles. Four children take one card from each set to create a 4-digit number. The children, in pairs, round the number to its nearest ten, hundred and thousand, scoring one point for each correct answer. Repeat for different 4-digit numbers. Continue until all the place-value cards have been used. The winner is the pair with the most points.

ACTIVITY 2
2 pairs

- *Rounding a 2-place decimal number to the nearest whole number*

Three dice (1 to 6, different colours), post-it notes, counters

The children decide which dice is the units, which is the tenths and which the hundredths. Each pair throws the three dice together and writes the decimal number they make on a post-it note. The two pairs swap numbers and round their new number to the nearest whole number. If it rounds to an even number, they take two counters; if it rounds to an odd number, they take one counter. They continue until one pair has ten counters.

ACTIVITY 3
3–4 children

- *Rounding a 2-place decimal number to the nearest whole number*

Post-it notes, counters

One child writes a whole number between 0 and 10 on a sheet of paper, e.g. 5. Each child writes on a post-it note a 2-place decimal number which will round up or down to the target number on the paper, e.g. 4·68 or 5·31. The children stick the post-it notes on the paper. They check each other's answers to make sure that all the numbers round to the target number. If they are correct, they take a counter. Repeat several times. Who collects the most counters?

ACTIVITY 4
2–3 children

- *Rounding a 2-place decimal number to the nearest whole number*
- *Rounding a 2-place decimal number to the nearest tenth*

Number cards (0 to 9) (PCM 9)

Spread out the cards, face up. Using these cards, the children make as many 2-place decimal numbers as possible which round to 5. Are they sure that they have found them all? How many can they make which round, to their nearest tenth, to 3·8?

ACTIVITY 5
3 children

- *Rounding a 5-digit number to its nearest thousand*

Game 1: 'Iceberg Ahead!', coloured cubes or counters

(See instructions on the card)

GROUP ACTIVITY 6
3 children

- *Creating a 1- or 2-place decimal number which rounds to a given whole number*

Place-value cards (U, t, h) (PCMs 3 to 6), a 10-sided dice (0 to 9)

GROUP ACTIVITY 7
3–4 children

- *Rounding a 2-place decimal number to the nearest tenth*

Two sets of number cards (0 to 9) (PCM 9), cubes

N1

Round to a dice

Place-value
GROUP ACTIVITY 6

3 children

> Place-value cards (U, t, h), a 10-sided dice (0 to 9)

Spread out the cards face up.

Take turns to roll the dice.

Choose cards to make a number which rounds to the dice number.

For example, if you roll a 3, you might make 2·9 or 2·96 with the cards.

Write down the card number and the dice throw.

Continue taking turns until everyone has written at least six numbers.

N1

Cloud nine

Place-value
GROUP ACTIVITY 7

3–4 children

> Two sets of numbers cards (0 to 9), interlocking cubes

Each draw a large grid like this:

U	t	h
·		

Shuffle the cards and place them face down in a pile.

Take turns to take a card. Place it in a space on your grid.

Take another card each, and place it in another space.

Take another card each and place it in the last space.

Each round your number to the nearest tenth.

Look at your answer. If your number rounds to one of the cloud numbers, take a cube.

Play again.

Abacus Ginn and Company 2001 Copying permitted for purchasing school only. This material is not copyright free.

N2 Place-value

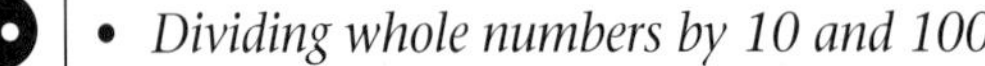

ACTIVITY 1
Whole class, in pairs

- *Dividing whole numbers by 10 and 100*

Place-value grid (Th, H, T, U) drawn on the board, two sets of number cards (1 to 9 and several 0 cards) (PCM 9), Blu-tack

One child chooses a card and holds it in the hundreds column. Another holds another in the tens column. All the children write the number, e.g. 340, and divide it by 10, i.e. they write '340 ÷ 10 = 34'. A third child demonstrates the answer by sliding the cards one place to the right. Repeat, dividing different 3- and 4-digit numbers by 10. Extend to dividing different numbers by 100.

ACTIVITY 2
3 children

- *Dividing a 2-digit number by 10*

Rulers (marked in cm and mm), assorted nuts and bolts, screws, paperclips and small metal objects, post-it notes

Each child chooses an object. They measure it and write its length in millimetres on a post-it note, e.g. 'bolt: 43 mm'. They pass on the object and note to another who divides the measurement by 10 and writes the answer on the note. This is the length in centimetres, e.g. 4·3 cm. They pass on the object and the note for the third child to check the measurement in centimetres. Repeat with the children choosing another object.

ACTIVITY 3
2–3 children

- *Dividing a 2-digit number by 10, calculating an average (mean)*

Six or seven books, post-it notes

Each child chooses a book and a page in it. They count the number of words in ten lines, write it on a post-it note and stick it on the page. They pass the book to another child to check the word count. They pass the book on again, and divide the number of words by 10 to estimate the average (mean) number of words in a line. They write the answer on the post-it note. They then count the words on one line. *How close is the estimate?* They open each book at another page and count the words on one line. *How close is the estimate?* Repeat with each child choosing a different book.

ACTIVITY 4
3–4 children

- *Dividing any whole number by 100*

Calculator, interlocking cubes

One child writes any number, e.g. '4560'. The other children divide it by 100 (without using the calculator), and write down the answer. The first child checks the answer on the calculator. The children who have written down a correct answer take a cube. Repeat with the children taking turns to create a number. Who collects the most cubes?

GROUP ACTIVITY 5
3 children

- *Dividing a 2-digit number by 10*

Place-value cards (T, U) (PCMs 1 to 4), interlocking cubes

GROUP ACTIVITY 6
3 children

- *Dividing a 3-digit number by 100*
- *Comparing the nearness of different decimal numbers by finding the difference between pairs of them*

Number cards (0 to 9) (PCM 9), interlocking cubes

What's my number?

Place-value
GROUP ACTIVITY 5

3 children

Place-value cards (T, U), interlocking cubes

Shuffle the cards separately and place them face down in two piles.

Take turns to take a card from each pile and secretly make a 2-digit number.

Divide the number by 10 and write the answer.

The others look at the answer and say what your number is. They write it down.

Compare with the card numbers.

They take a cube for each correct answer.

Play again, until all the cards are taken.

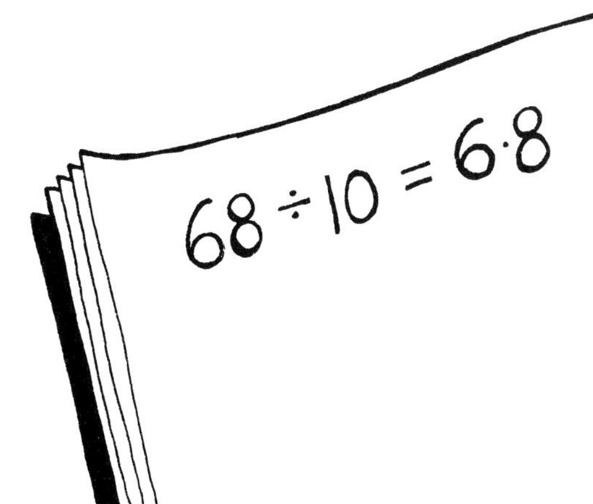

Close numbers

Place-value
GROUP ACTIVITY 6

3 children

Number cards (0 to 9), interlocking cubes

Draw a large copy of this grid:

H	T	U

Shuffle the cards and place them face down in a pile.

Each think of a decimal number, with tenths and hundredths. Write it down.

Take one card each and place them on the grid, to make a 3-digit number.

Read the number aloud.

Divide it by 100 and write down the answer. Check that you all agree.

Compare the answer with your decimal numbers. Whoever is closest takes a cube.

Return the cards to the bottom of the pile and play again.

The winner is the first to collect six cubes.

Abacus Ginn and Company 2001 Copying permitted for purchasing school only. This material is not copyright free.

N3 Multiplication/division

ACTIVITY 1
Whole class, in pairs

- *Recognising numbers up to 10 which will divide into a given number*

Number cards (1 to 10) (PCM 9), two dice (1 to 6), counters

Choose a child to throw both dice and create a 2-digit number. Write it on the board. The children, in pairs, lay out every number card which will divide into the 2-digit number. Discuss the correct answers, giving one counter for each correct number, and removing one counter for each incorrect answer. Repeat several times. Which pair collects the most counters?

ACTIVITY 2
3 children

- *Multiplying a 2-digit number by a 1-digit number*

Number cards (0 to 9) (PCM 9), a dice (1 to 6)

Shuffle the cards and place them in a pile, face down. The children take turns to throw the dice and take a card. They multiply the two numbers together and write down the answer. They check each other's work. Repeat several times.

ACTIVITY 3
2 pairs

- *Recalling division facts*

Number cards (1 to 10) (PCM 9)

Shuffle the cards and place them in a pile face down. One pair takes two cards. They secretly multiply the numbers together and write down the answer. They show the answer and one of the cards to the other pair. The second pair works out the number on the unseen card and writes down the multiplication. Repeat, swapping roles.

ACTIVITY 4
2–3 children

- *Recalling multiplication facts, investigating all possible products when multiplying two different numbers up to 10*

Number cards (0 to 9) (PCM 9)

The children choose pairs of cards to make multiplications. How many different answers can they make? Remind them that they cannot use the same number card twice, e.g. 3×3.

ACTIVITY 5
3 children

- *Recalling multiplication facts*

Game 2: 'Snaky tables', number cards (3 to 9) (PCM 9), two dice (1 to 6), counters (different colours for each player)

(See instructions on the card.)

GROUP ACTIVITY 6
3 children

- *Recalling multiplication facts*

Three sets of number cards (0 to 9) (PCM 9), post-it notes, interlocking cubes

GROUP ACTIVITY 7
2–3 children

- *Searching for different pairs of factors of a number*

Two sets of number cards (0 to 10) (PCM 9), post-it notes

N3

Grid times

Multiplication/division
GROUP ACTIVITY 6

3 children

Three sets of number cards (0 to 9), post-it notes, interlocking cubes

Shuffle the cards and place them face down in a pile.

Each draw a large grid like this:

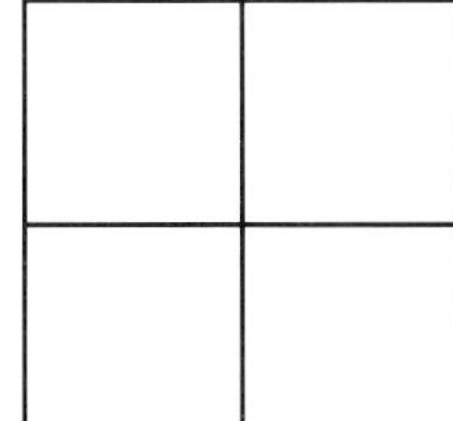

Take turns to take four cards each, and place them on your grid.

Multiply the pairs of numbers down and across.

Write the four answers on stickers and place them around your grid.

The player with the largest answer takes two cubes.

The player with the next largest answer takes one cube.

Return the cards to the bottom of the pile and repeat the activity.

Continue until one player has five cubes.

N3

Cloud multiplying

Multiplication/division
GROUP ACTIVITY 7

2–3 children

Two sets of number cards (0 to 10), post-it notes

Try to place four different cards to make the cloud multiplication correct.

For example, $2 \times 10 = 20 = 4 \times 5$.

Each time, write the answer on a post-it note and put it in the cloud.

Copy the whole cloud multiplication and start again.

How many different cloud multiplications can you write?

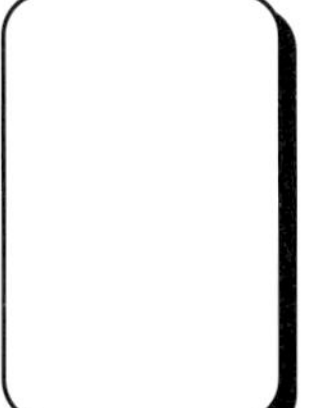    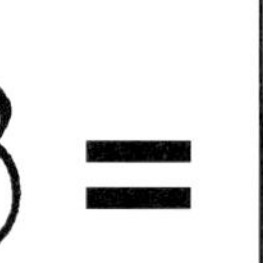   

Abacus Ginn and Company 2001 Copying permitted for purchasing school only. This material is not copyright free.

N4 Multiplication/division

ACTIVITY 1
Whole class, in pairs

• *Doubling 3-digit multiples of 10*

Place-value cards (H, T) (PCMs 1, 2)

Two children select one card of each type to create a 3-digit number, e.g. 460. The children double the number and write it down. They can try this in their heads first, then double the hundreds and tens and combine the two, e.g. double 460 = double 400 (800) + double 60 (120) = 920. Discuss the answer together. Give one point for each correct answer. Repeat several times, using different card numbers. Which pair scores the most points?

ACTIVITY 2
2 pairs

• *Doubling a number up to 66*

A set of dominoes, a dice (1 to 6), counters

Lay out the dominoes face down. Each pair chooses a domino and arranges it to make a 2-digit number, e.g. domino 3/4 can make either 34 or 43. They write down the number and the other pair doubles the number and writes it down. The children check each other's calculation. If they are correct they collect a counter. Continue until each pair has played ten rounds. Which pair collects the most counters?

ACTIVITY 3
2 pairs

• *Halving and doubling amounts of money*

A cloth bag, coins (£1, 10p), counters

Place the coins in the bag. One pair secretly takes an amount from the bag, and counts it, e.g. £4·60. They calculate double this amount and write it down on a piece of paper, i.e. £9·20, giving the paper to the other pair. The other pair must halve this amount to find out how much money was taken from the bag. They write down half of the amount and then count the coins to see if they are correct. Repeat several times with the children swapping roles.

ACTIVITY 4
2–3 children

• *Doubling 1-place decimal numbers*

Number cards (0 to 9) (PCM 9), counters

The children create two pairs of 1-place decimal numbers with the cards, using the counters as decimal points. One pair of numbers must be double the other. For example, they can create the pair 6·4 and 12·8, but not 14·2 and 28·4 because there is only one card for each digit. Let them investigate how many different pairs they can create which have one number more than 10. Extend the activity to creating pairs of 2-place decimal numbers.

GROUP ACTIVITY 5
2 pairs

• *Doubling and halving 1-place decimal numbers*

Place-value cards (U, t) (PCMs 3 to 5), counters

GROUP ACTIVITY 6
2–3 children

• *Halving even numbers up to 200*

Counters

Double the number

Multiplication/division
GROUP ACTIVITY 5

Place-value cards (U, t), counters

2 pairs

In pairs, take turns to make a number like this using the place-value cards:

Then the other pair says:

1. the number *Seven point four.*

2. its double. *Fourteen point eight.*

If they are correct, they collect a counter.

Play ten rounds. Who collects the most counters?

Play again, but only make numbers with an even tenths digit. This time, halve the number instead of doubling it.

Halve the number

Multiplication/division
GROUP ACTIVITY 6

2–3 children

Cover each number with a counter.

Take turns to remove a counter. Say half of the number.

If you are correct, keep the counter. Check each other's answers.

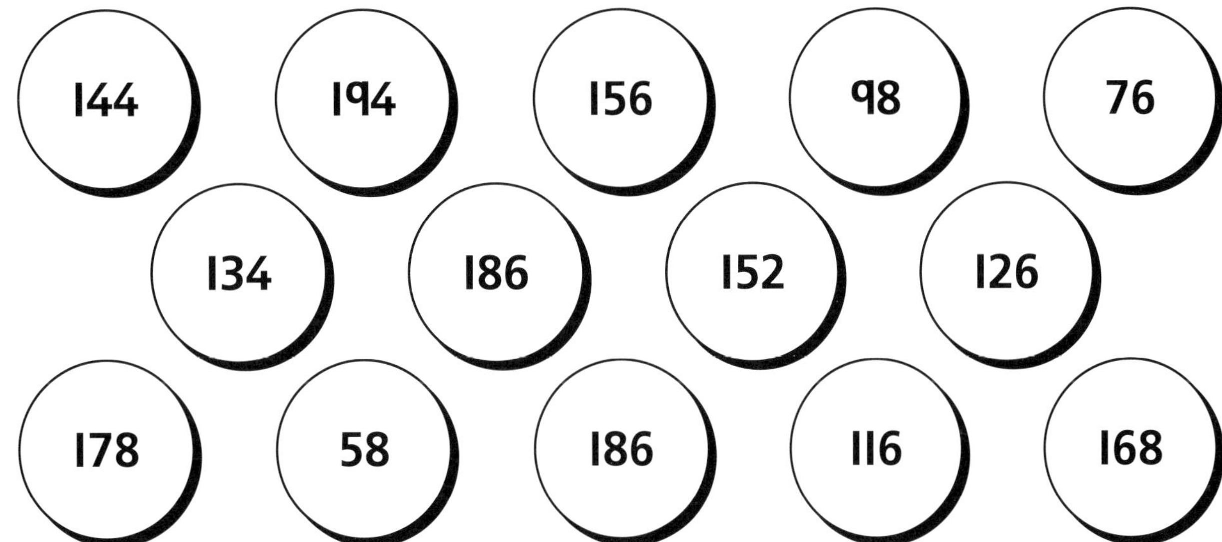

Abacus Ginn and Company 2001 Copying permitted for purchasing school only. This material is not copyright free.

N5 Multiplication/division

ACTIVITY 1
Whole class, in pairs

- *Multiplying a 2-digit number by 99, by multiplying by 100, then adjusting*

Number cards (10 to 30) (PCMs 9 to 11), a calculator

Select one card at random, e.g. 23. The children multiply the number by 99, by first multiplying it by 100 and then adjusting the answer. One child checks the answer with the calculator. Pairs score one point for a correct answer. Repeat ten times. Who scored ten points, nine points, ...?

ACTIVITY 2
2–3 children

- *Investigating patterns in multiplying a 2-digit number by 101*

Number cards (10 to 99) (PCMs 9 to 15), calculators

Shuffle the cards, and place them in a pile, face down. One child reveals the top card and all the children multiply it by 101, using calculators. They write down the multiplication, e.g. '34 × 101 = 3434'. Repeat several times, with different number cards. The children look for a pattern in the numbers to find a quick way of multiplying a 2-digit number by 101.

ACTIVITY 3
3 children

- *Multiplying by 50 by multiplying by 100, then halving*

Place-value cards (T, U) (PCMs 1 to 4)

Lay the cards face down. Each child selects three tens cards and three units cards to make three 2-digit numbers. They place all nine numbers in a column. Each child writes them down in a column. They then multiply each number by 50, by first multiplying each one by 100, then halving. The children check each other's answers and score one point for each correct answer. Who scored the most points?

ACTIVITY 4
2–3 children

- *Multiplying by near multiples of 50, by multiplying by 50, then adjusting*

Number cards (10 to 50) (PCMs 9 to 12)

The children select a card at random. They multiply it by 49, then by 51, by multiplying by 50 and then adjusting. They then find the difference between the two answers. Repeat with different number cards. The children look for patterns in the answers, and for quick ways of doing the calculation. They then investigate quick ways of finding the difference between the answers resulting from multiplying a number by both 48 and 52, and 46 and 54.

GROUP ACTIVITY 5
2 pairs

- *Multiplying by near multiples of 50, by multiplying by 50, then adjusting*

Counters (two colours), a calculator

GROUP ACTIVITY 6
2–3 children

- *Multiplying by near multiples of 100, by multiplying by 100, then adjusting*

Counters, interlocking cubes, a one-minute timer, a calculator

Three in a line

2 pairs

Counters (two colours), a calculator

In pairs, take turns to choose a number from the grid for the other pair.

Ask them to multiply this number by 51.

Check their answer with a calculator.

If they are correct, they cover the number with one of their counters.

If they are incorrect, you cover it with one of your counters.

The winning pair is the first to have three counters in a straight line.

Play again, multiplying the numbers by 49.

17	41	38	25
32	27	47	53
16	35	19	51
57	22	46	62

Multiplication frenzy

2–3 children

Counters, interlocking cubes, a one-minute timer, a calculator

Cover each oval with a counter.

Take turns to remove a counter. You each have one minute, using the timer, to do the calculation. Check the results together, using a calculator if necessary.

If you are correct, collect a cube.

When all the counters have been removed, the winner is the player with the most cubes.

101×32 56×99 99×46

18×101 99×28 75×99 101×38

43×101 102×56 101×25 98×73

Abacus Ginn and Company 2001 Copying permitted for purchasing school only. This material is not copyright free.

N6 Multiplication/division

ACTIVITY 1
Whole class, in pairs

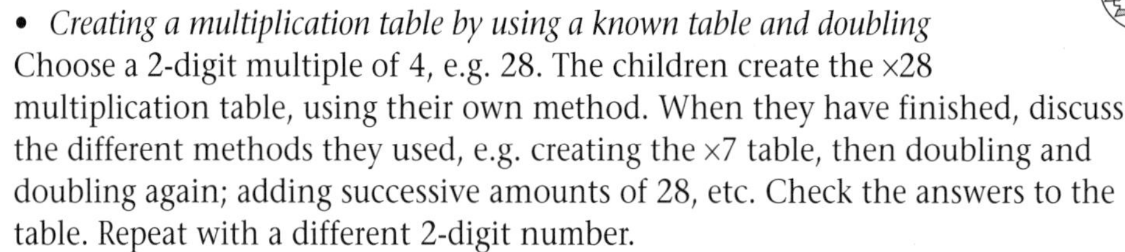

• *Creating a multiplication table by using a known table and doubling*
Choose a 2-digit multiple of 4, e.g. 28. The children create the ×28 multiplication table, using their own method. When they have finished, discuss the different methods they used, e.g. creating the ×7 table, then doubling and doubling again; adding successive amounts of 28, etc. Check the answers to the table. Repeat with a different 2-digit number.

ACTIVITY 2
2–3 children

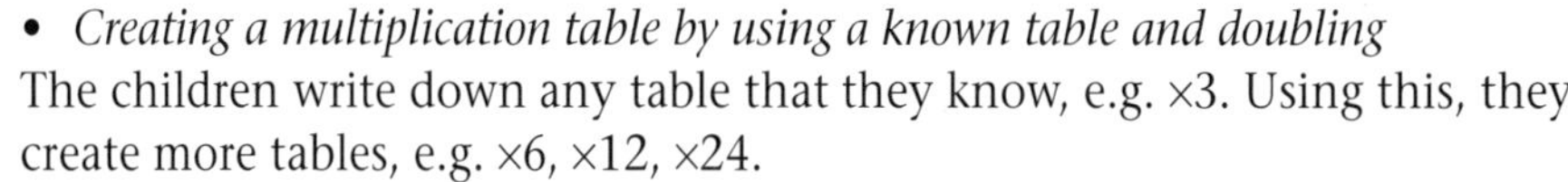

• *Creating a multiplication table by using a known table and doubling*
The children write down any table that they know, e.g. ×3. Using this, they create more tables, e.g. ×6, ×12, ×24.

ACTIVITY 3
3 children

• *Multiplying by 25, by multiplying by 100, then halving and halving again*
Place-value cards (T, U) (PCMs 1 to 4)
Lay out the cards face down. Each child selects three tens cards and three units cards to make three 2-digit numbers. They put the nine numbers together in a column. All the children write them down in a column. They then multiply each number by 25, first by multiplying each one by 100, then halving, then halving again. The children check each other's answers, scoring one point for each correct answer. Who scored the most points?

ACTIVITY 4
2–3 children

• *Investigating the creation of unknown multiplication facts from one known fact*
The children start with any multiplication fact involving even numbers, e.g. 16 × 36 = 576. They use this to see how many other multiplication facts they can create, without doing any multiplying, e.g. 8 × 36, 16 × 18, 32 × 36.

ACTIVITY 5
Pairs

• *Estimating the result of a multiplication by 25*
Spreadsheet software
Prepare a spreadsheet file similar to the one shown. Use the 'If logical' function **=IF(E2>0,E2–(A2*C2)," ")** in Column F. The children estimate the result of multiplying the numbers in Column A by 25 and enter their estimates in Column E. They check Column F to see how near their estimates were to the actual answers.

	A	B	C	D	E	F
1	Number	multiplied by	25	approximately equals	Estimate	How close were you?
2	14	×	25	≈	300	-50
3	88	×	25	≈	2400	200
4	94	×	25	≈		
5	16	×	25	≈		

GROUP ACTIVITY 6
3–4 children

• *Creating a multiplication table by using a known table and doubling*

GROUP ACTIVITY 7
3–4 children

• *Multiplying by 25, by multiplying by 100, then halving and halving again, estimating the result of a multiplication by 25*

Table tops

Multiplication/division
GROUP ACTIVITY 6

3–4 children

Copy and complete each table.

Use the results of the ×8 table to write the ×16 table and the ×32 table.

Use the results of the ×9 table to write the ×18 table and the ×36 table.

1 × 8 = 8	1 × 9 = 9
2 × 8 = 16	2 × 9 = 18
3 × 8 = 24	3 × 9 = 27
4 × 8 =	4 × 9 =
5 × 8 =	5 × 9 =
6 × 8 =	6 × 9 =
7 × 8 =	7 × 9 =
8 × 8 =	8 × 9 =
9 × 8 =	9 × 9 =
10 × 8 =	10 × 9 =
11 × 8 =	11 × 9 =
12 × 8 =	12 × 9 =

Estimate the product

Multiplication/division
GROUP ACTIVITY 7

3–4 children

Copy the scoresheet.

Start by writing an estimate for each product.

Complete the multiplication by first multiplying by 100, then halving your answer, then halving it again.

How close are your estimates?

Round	Multiplication	Estimate	× 100	× 50	× 25
1	14 × 25				
2	37 × 25				
3	52 × 25				
4	25 × 73				
5	25 × 46				
6	86 × 25				
7	25 × 94				
8	67 × 25				

Abacus Ginn and Company 2001 Copying permitted for purchasing school only. This material is not copyright free.

Fractions/decimals

ACTIVITY 1
Whole class, in pairs

- *Finding fractions of quantities*

Coins (£1, £2, 50p, 20p, 10p, 5p, 2p, 1p), a cloth bag

One child in each pair takes a large handful of coins from the bag. The other child writes down a fraction less than 1 with a numerator greater than 1, e.g. $\frac{3}{4}$. The first child counts the money, e.g. £2·56. Both children then write down the amount and work out the given fraction of that amount to the nearest penny. They compare their calculations, agree on a correct answer and write it down, i.e. $\frac{3}{4}$ of £2·56 = 64p. Repeat ten times with the children swapping roles.

ACTIVITY 2
3–4 children

- *Finding fractions of quantities*

Commercial video tapes, fraction cards ($\frac{2}{3}$, $\frac{3}{4}$, $\frac{3}{5}$, $\frac{4}{7}$, $\frac{3}{8}$, $\frac{5}{9}$, $\frac{7}{10}$) (PCM 19)

The children select one of the videos and each writes down its running time. They spread out the fraction cards face down. Each child takes a card and works out how many minutes of the video they will watch if they watch that fraction of the whole. For example, if the running time of the video is 95 minutes and the fraction card is $\frac{4}{7}$, then one seventh is about $13\frac{1}{2}$ minutes so $\frac{4}{7}$ is about 54 minutes. The children compare their calculations. The children with the longest and the shortest times each score five points. They replace the fraction cards and take a different video. Repeat until someone has scored 30 points.

ACTIVITY 3
2 pairs

- *Finding fractions of quantities*

Several books, number cards (2 to 9) (PCM 9)

Shuffle the cards and place them face down in a pile. One child takes two cards and creates a fraction less than 1. The other child takes a book and looks at the total number of pages. Together, the children calculate how many pages of the book they would read if they read that fraction. They agree on a correct answer and write down the calculation, e.g. $\frac{3}{4}$ of 164 = 123. Repeat the process ten times.

ACTIVITY 4
2 pairs

- *Finding fractions of quantities*

Interlocking cubes, a cloth bag, fraction cards ($\frac{2}{5}$, $\frac{2}{3}$, $\frac{3}{4}$, $\frac{5}{6}$, $\frac{7}{8}$) (PCM 19)

One child takes a large handful of cubes from the bag. The other child chooses a fraction card, e.g. $\frac{3}{4}$. The first child counts the number of cubes, e.g. 25 and they both work out the given fraction. If they cannot exactly divide the denominator into their number, they should get rid of, or take, one or two cubes, i.e. they get rid of one cube, leaving 24. They agree on a correct answer and write down the matching calculation, i.e. $\frac{3}{4}$ of 24 = 18. Repeat ten times.

GROUP ACTIVITY 5
3–4 children

- *Finding fractions of quantities*

Counters, interlocking cubes

GROUP ACTIVITY 6
4 children

- *Finding fractions of quantities*

Counters

N7

Fractional account

Fractions/decimals
GROUP ACTIVITY 5

3–4 children

Counters, interlocking cubes

Cover each fraction with a counter.

Take turns to remove a counter. Think of a number and work out that fraction of the number. (You may write some notes.) Say your answer out loud. The others have to tell you what your number is. For example, you take the counter from $\frac{3}{5}$ and think of 35. $\frac{3}{5}$ of 35 is 21. You say 21 and the others guess that the number is 35.

If your calculation was correct, take a cube. Play until one of you has five cubes.

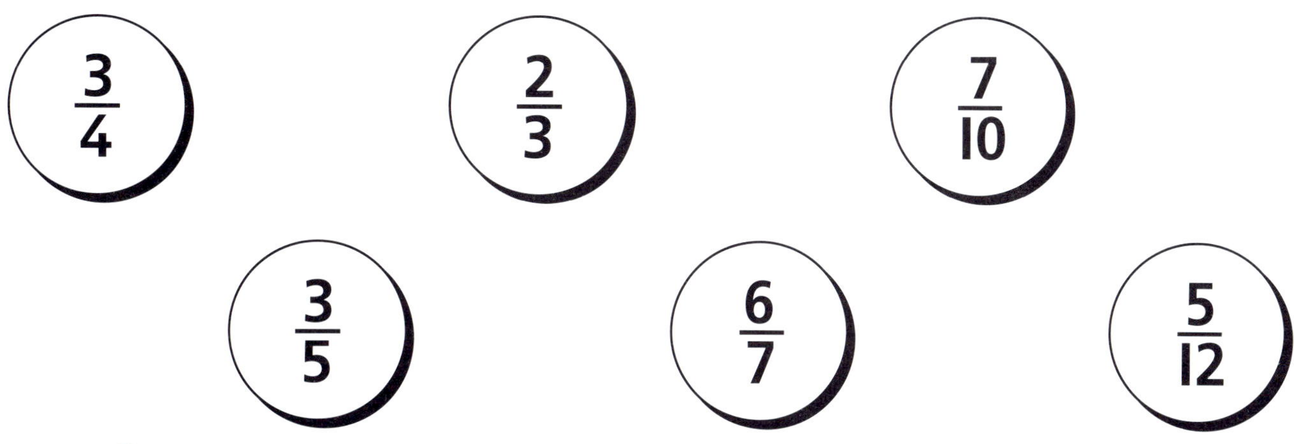

N7

Swinging sixties

Fractions/decimals
GROUP ACTIVITY 6

4 children

Counters

Cover each fish with a counter.

Each remove a counter and work out that fraction of 60.

Check each other's answers. Each player with a correct answer scores ten points. Keep playing until all the counters have been removed. Who has the most points?

Play again, this time finding the fractions of 90.

Abacus Ginn and Company 2001 Copying permitted for purchasing school only. This material is not copyright free.

N8 Fractions/decimals

ACTIVITY 1
Whole class, in pairs

- *Writing fractions in their simplest form*

Multiplication square (1 to 100) (PCM 17), post-it notes

One child secretly chooses two rows on the grid which are not next to each other. On a post-it note, they write a fraction using a number from each of the chosen rows, e.g. $\frac{12}{40}$. The other child then points to the two chosen rows and writes the fraction in its simplest form, on the same post-it. Repeat ten times with the children swapping roles.

ACTIVITY 2
4 children

- *Creating pairs of equivalent fractions*

Three sets of number cards (2, 3, 4, 5, 6, 8, 9, 10, 12, 15, 18, 20) (PCMs 9, 10)

Shuffle the cards and deal out five to each child, leaving the rest in a pile. One child puts down a card, face up. A second child puts down a card above or below it to make a fraction, or next to it to start a new fraction. The other two children do the same. However, at any time, there may only be three cards on the table which are not part of fractions. Whenever a card is put down, another card must be picked up from the pile. The aim is to create pairs of equivalent fractions, e.g. $\frac{3}{4}$ and $\frac{6}{8}$. All the fractions created must be less than one. The child who places the last card to make a pair of equivalent fractions scores five points The winner is the child with the most points when all the cards have been played.

ACTIVITY 3
3 children

- *Finding pairs of equivalent fractions*

Interlocking cubes

Each child builds a row of 24 cubes. One child divides their row into halves, the second child divides their row into quarters, the third child divides their row into eighths. The children compare their fractions. *How many quarters or eighths are equivalent to one half? How many eighths are equivalent to one quarter? To three quarters?* The children write down all the pairs of equivalent fractions they can find, e.g. $\frac{3}{4} = \frac{6}{8}$. Repeat, building rows of 18 and dividing the rows into halves, thirds and sixths, and also building rows of 36 and dividing into halves, thirds, sixths and twelfths.

GROUP ACTIVITY 4
3–4 children

- *Writing fractions in their simplest form*

Counters, interlocking cubes

GROUP ACTIVITY 5
3–4 children

- *Finding equivalent fractions*

Small blank cards

Counter moves

Fractions/decimals
GROUP ACTIVITY 4

3–4 children

Counters, interlocking cubes

Cover each balloon with a counter.

Take turns to play. Remove a counter. Write the simplest equivalent fraction to the fraction on the balloon. If the others agree that it is correct, take a cube.

Play until all the fractions are uncovered.

Replace the counters. Play again, but this time write a different equivalent fraction.

Card fractions

Fractions/decimals
GROUP ACTIVITY 5

3–4 children

Small blank cards

Write different fractions on the cards. Spread the cards out face down.

Take turns to turn over a card. Write an equivalent fraction on the back of the card. The others check that it is correct. Continue until all the cards have fractions on both sides.

Take turns to point at a card and say the fraction on the hidden side. Turn it over to check that you are correct. If you are correct, keep the card. If not, replace it. Play until all the cards are taken. Who has the most cards?

Abacus Ginn and Company 2001 Copying permitted for purchasing school only. This material is not copyright free.

N9 Fractions/decimals

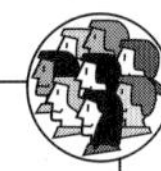

ACTIVITY 1
Whole class, in pairs

- *Giving fractions a common denominator and comparing fractions*

A dice (1 to 6)

Each child writes a fraction less than 1 in its simplest form. The children throw the dice twice. They write the smaller number over the larger one to create a fraction. The child whose own fraction is closest to the one created by the roll of the dice wins. To work out the closest fraction, the children must make the denominators of their fractions the same. They decide on a common denominator, convert their fractions and then compare them to see who has won. Repeat six times, writing new fractions every time.

ACTIVITY 2
3 children

- *Giving fractions a common denominator and comparing fractions*

Number cards (2 to 9) (PCM 9)

Shuffle the cards and place them face down in a pile. Each child takes two cards and creates a fraction less than 1 by putting the smaller number over the larger. If necessary, they reduce the fraction to its simplest form. The children compare their three fractions. The child whose fraction is closest to one half wins. To work this out, the children may have to make the denominators of the fractions the same. The children decide on a common denominator and convert their fractions. They then compare their fractions to see which is closest to one half. The children replace the cards at the bottom of the pile. Repeat six times.

ACTIVITY 3
3 children

- *Converting fractions into twelfths*

Fraction cards ($\frac{1}{4}$, $\frac{1}{3}$, $\frac{1}{2}$, $\frac{2}{3}$, $\frac{3}{4}$, $\frac{1}{6}$, $\frac{5}{6}$) (PCM 19), a number line (0 to 1, marked in twelfths)

Shuffle the cards and place them in a pile face down. The children each take a fraction card and compare them. They must place each card on the number line in the correct position. They discuss where each card should go. This will help them to convert their fraction into twelfths. Each child then writes the equivalent fractions, e.g. $\frac{3}{4} = \frac{9}{12}$. The children take a new card each and repeat the activity until all the cards are gone. *Which fractions along the line have no card?* The children should write those fractions.

ACTIVITY 4
3 children

- *Ordering fractions*

Number cards (1 to 12) (PCMs 9, 10), a number line (0 to 1)

The children use the cards to create six fractions. They arrange them in order along the number line in approximately the correct place. The children then draw the number line with the fractions. Repeat with different fractions. How many different lines with six fractions can they draw?

GROUP ACTIVITY 5
3–4 children

- *Giving fractions a common denominator and comparing fractions*

Counters, interlocking cubes

GROUP ACTIVITY 6
3–4 children

- *Finding the mid-point between a pair of fractions*

Post-it notes

Snail trail

Fractions/decimals
GROUP ACTIVITY 5

3–4 children

Counters, interlocking cubes

Cover each snail with a counter.

Each write a fraction that is less than 1 on a piece of paper.

Remove one counter. Read the fraction that is uncovered.

Convert your fractions so that they have a common denominator with the uncovered fraction.

Compare your fractions. If your fraction is closest to the uncovered one, you win and take a cube. If your fraction is the same, you are disqualified.

Replace the counter. Play again, but this time write down a different fraction.

Half-way house

Fractions/decimals
GROUP ACTIVITY 6

3–4 children

Take turns to choose a pair of fractions and work out the fraction half-way between them.

Check each other's answers. Use the number line to help.

Copy the number line and mark the two chosen fractions and the half-way fraction on it.

Continue until all the fractions have been marked on the number line.

$\frac{1}{8}$ $\frac{1}{4}$

$\frac{1}{6}$ $\frac{1}{3}$

$\frac{1}{10}$ $\frac{1}{5}$

$\frac{1}{6}$ $\frac{1}{12}$

$\frac{1}{2}$ $\frac{1}{4}$

$\frac{2}{3}$ $\frac{5}{6}$

$\frac{2}{7}$ $\frac{3}{7}$

$\frac{3}{4}$ 1

$\frac{4}{9}$ $\frac{5}{9}$

$\frac{4}{5}$ $\frac{9}{10}$

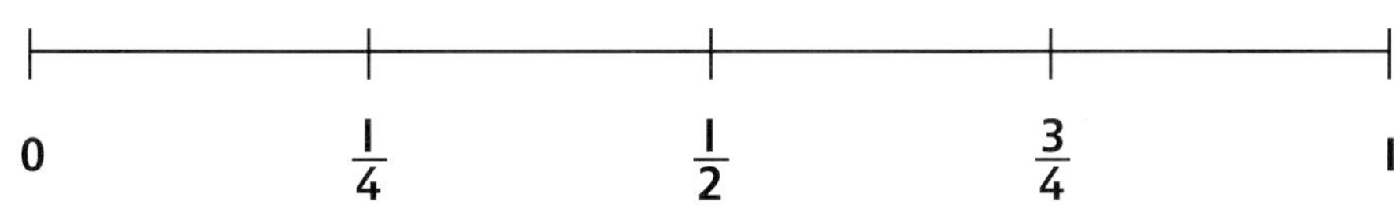

Abacus Ginn and Company 2001 Copying permitted for purchasing school only. This material is not copyright free.

N10 Addition/subtraction

ACTIVITY 1
Whole class, in groups of 3

• *Calculating amounts of money by counting on*

Coins (ten 1p, nine 10p, nine £1), a cloth bag, post-it notes, interlocking cubes

The children count the money and place the coins in a bag. One child takes a handful of coins. The second child counts them and writes the amount on a post-it note. The third child works out how much is left in the bag and writes it on the note below the first amount. The first child now counts the amount left in the bag. If the second child was correct, they take a cube. Repeat eleven more times, with the children swapping roles. Which group has the most cubes?

ACTIVITY 2
3 children

• *Counting on from a 3-digit number to make 1000*

Number cards (0 to 9) (PCM 9), a calculator, counters

The children decide who will be 'hundreds', who will be 'tens' and who will be 'units'. They each take one card and place it in their chosen position to make a 3-digit number. They each write down what must be added to the number to make 1000. The children compare their answers and use a calculator to check. Those who are correct take a counter. The children return the cards and repeat ten times.

ACTIVITY 3
3 children

• *Calculating amounts of money by counting on*

Number cards (1 to 100) (PCMs 9 to 15), coins (10p, 1p), post-it notes

One child takes a card. The second child matches the number with the same number of pence, in 10p and 1p coins. The third child works out how much must be added to make £1. The children agree on the amount, write it on a post-it note and stick it on the card. Repeat nine more times with the children swapping roles.

ACTIVITY 4
Pairs

• *Adding two 3-digit numbers to make 1000*

Number cards (0 to 9) (PCM 9)

The children spread out the cards face up. They use six cards to make two 3-digit numbers that together total 1000. Can they make ten different additions?

GROUP ACTIVITY 5
4 children

• *Adding money to make a given amount*

GROUP ACTIVITY 6
4 children

• *Finding pairs of numbers that total 100*

Number grid (1 to 100) (PCM 18), counters (four colours)

Purse pointer

Addition/subtraction
GROUP ACTIVITY 5

4 children

Each choose a purse. Take turns to choose a card. Add the amount on the card to the amount in your purse. Write down the total.

Continue choosing cards, keeping a running total.

Keep playing until one of you has exactly £5.
If you make over £5 you are out.

Play again, this time choosing a different purse.

Four in a line

Addition/subtraction
GROUP ACTIVITY 6

4 children

Number grid (1 to 100), counters (four colours)

Take turns to play. Place two counters on any two squares that add up to 100, e.g. 34 and 66.

Check each other's totals.

Continue playing like this until one of you has four counters in a straight line in any direction. That person scores 10 points.

Remove all the counters from the grid and play again. Repeat until one of you has scored 50 points.

1	2	3	4	5	6	7	8	8	10
11	12	13	14	15	16	17	18	19	20
21	22	23	24		26	27	28	29	30
31	32	33		35	36	37	38	39	40
41		43	44	45	46	47	48	49	50
51	52	53	54	55	56	57		59	60
61	62	63	64	65		67	68	69	70
71	72	73	74		76	77	78	79	80
81	82	83	84	85	86	87	88	89	90
91	92	93	94	95	96	97	98	99	100

Abacus Ginn and Company 2001 Copying permitted for purchasing school only. This material is not copyright free.

N11 Addition/subtraction

ACTIVITY 1
Whole class, in pairs

- *Making 100 with three 1- or 2-digit numbers*

Number cards (1 to 100) (PCMs 9 to 15), post-it notes
The children deal out five cards each. They each add three of their cards to make a total as close as possible to 100. They lay them out face up, write the total on a post-it note and stick the note close to the cards. They score the difference between their total and 100. Repeat until one child's score goes over 100. The child with the lowest score wins. Play again.

ACTIVITY 2
3 children

- *Adding several 2-digit numbers*

Two dice (1 to 6)
The children take turns to throw two dice and create a 2-digit number. They write it down on a score sheet and continue, keeping a running total of their score. For example, if they throw a 3 and a 4, they can make 43. On their next turn, they throw a 2 and a 6 and make 62, add this to 43 and make 105. The children check each other's scores. Repeat until all three scores are over 300. The child with the highest score wins. Play again.

ACTIVITY 3
2 pairs

- *Creating and solving a number problem*

One child writes an addition using three different 2- or 3-digit numbers. The other child writes a story to turn the addition into a number problem. Each pair then gives their number problem to the other pair to solve. Repeat five times.

ACTIVITY 4
3–4 children

- *Adding several 2-digit numbers*

Game 3: 'Totals Tracker', counters, a dice (1 to 6)
(See instructions on the card.)

GROUP ACTIVITY 5
2 pairs

- *Adding three amounts of money*

GROUP ACTIVITY 6
2 pairs

- *Adding 2- and 3-digit numbers*

N11

Purse choice

Addition/subtraction
GROUP ACTIVITY 5

2 pairs

One pair chooses three purses.

That pair adds together the amount in each purse and writes down the total.

Show your total to the other pair and ask them to guess which three amounts you added together.

If they are correct, they score 10 points.

Play again, with the other pair choosing the purses and adding.

Repeat until each pair has chosen the purses six times.

N11

Make a crossword

Addition/subtraction
GROUP ACTIVITY 6

2 pairs

Use the grid below and fill the lines across and down with 2- or 3-digit numbers that fit exactly.

Write a set of clues for each number, such as two numbers which add together to make that number.

Make up an empty grid and a set of clues to give to the other pair to solve.

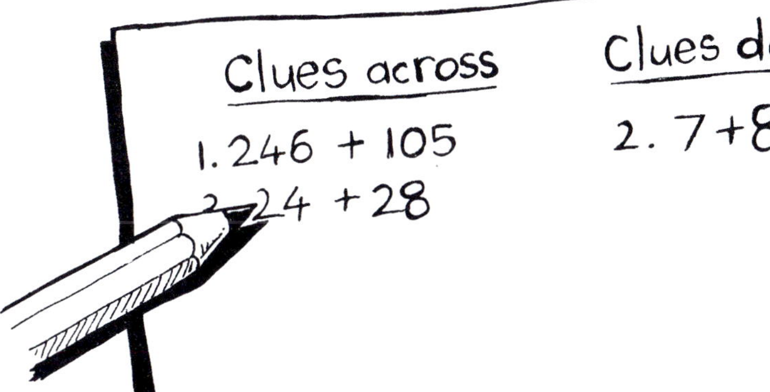

Abacus Ginn and Company 2001 Copying permitted for purchasing school only. This material is not copyright free.

N12 Addition/subtraction

ACTIVITY 1
Whole class, in pairs

• *Counting on from smaller to larger numbers*

A book with more than 150 pages, post-it notes

Each child in the pair chooses one page in the book and they both write down the page number on the same post-it note. They work out how many pages there are from one page to the next and write the difference on the note under the other two numbers. They check that they agree on the answer. Repeat ten times.

ACTIVITY 2
3–4 children

• *Repeatedly subtracting 2- and 3-digit numbers*

Three dice (1 to 6)

Each child writes down 1000 at the top of a sheet of paper. In turn, they throw two or three dice, create a 2- or 3-digit number and subtract it from 1000. They write down the new total and repeat until each child is left with a positive number as close as possible to zero. The child with the lowest remaining score wins. Repeat.

ACTIVITY 3
4 children

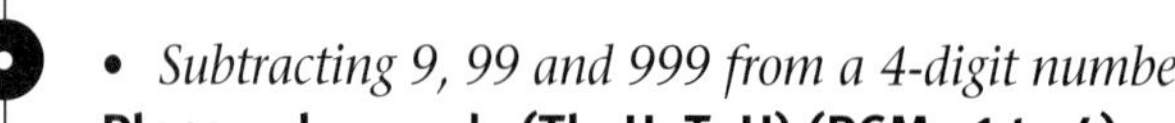

• *Subtracting 9, 99 and 999 from a 4-digit number*

Place-value cards (Th, H, T, U) (PCMs 1 to 4), post-it notes

Each set of cards is shuffled separately and placed in a pile, face down. The children take one card each and create a 4-digit number. They each write the number on the top of a post-it note. Below this, they write the number 9 less, below this the number 99 less than that and below this the number 999 less than that. The children compare each other's calculations. Do any of them need to alter their calculations? The children stick their post-it note to the number they created and repeat. They keep playing until all the cards have been used.

ACTIVITY 4
3–4 children

• *Counting on from 2-digit numbers to 100*

Game 4: 'Zigs and Zags', cubes, a dice (1 to 6), small pieces of card

(See instructions on the card.)

GROUP ACTIVITY 5
2 pairs

• *Subtracting two 4-digit numbers*

Place-value cards (Th, H, T, U) (PCMs 1 to 4)

GROUP ACTIVITY 6
3–4 children

• *Subtracting two 3-digit numbers*

Interlocking cubes

N12

Close to 5000

Addition/subtraction
GROUP ACTIVITY 5

2 pairs

Place-value cards (Th, H, T, U)

Shuffle the cards separately and place them in four piles face down.

Each pair takes two cards from each of the piles.

Use the cards to create two 4-digit numbers.

Subtract the smaller number from the larger, and write down the total.

Check each other's work.

The pair with the answer closest to 5000 wins.

Play again.

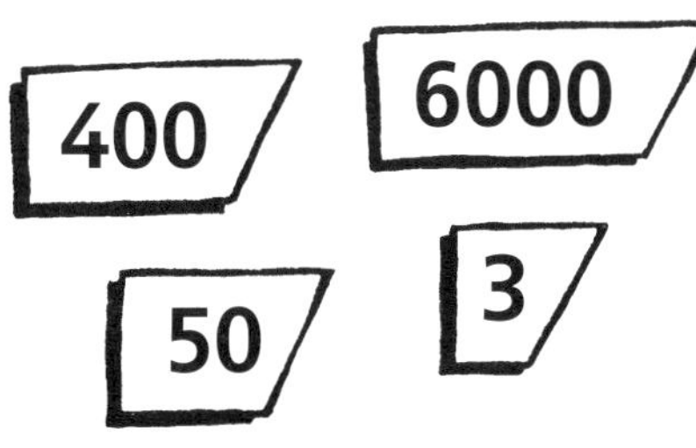

N12

Which clouds?

Addition/subtraction
GROUP ACTIVITY 6

3–4 children

Interlocking cubes

Take turns to play. Choose two cloud numbers and, in your head, work out the difference between them and write it down.

Show the other children your answer. The others have to tell you which two numbers you chose.

If they can all tell you the correct numbers, you can take a cube.

Play until you have each chosen five pairs of numbers. Who has the most cubes?

Abacus Ginn and Company 2001 Copying permitted for purchasing school only. This material is not copyright free.

N13 Properties of number

ACTIVITY 1
Whole class, in pairs

• *Finding the smallest common multiple of two numbers, up to 10*

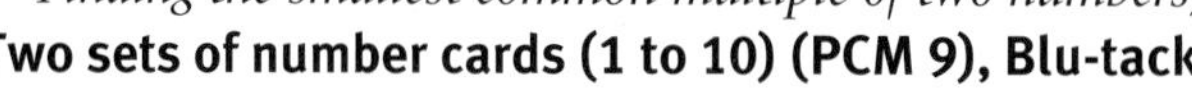

Two sets of number cards (1 to 10) (PCM 9), Blu-tack

The children choose pairs of cards at random and stick them on the board in pairs. There should be ten pairs of cards. Each pair writes the smallest common multiple of each pair of numbers. When they have done this, discuss their answers. The pairs score one point for each correct smallest common multiple. Which pair has the most points?

ACTIVITY 2
2–3 children

• *Constructing a list of the multiples of a given number*

One child chooses a number, e.g. 4, and writes down its first multiple, i.e. 4. They pass the paper to the next child who writes down the next multiple of 4, i.e. 8. The children continue up to the tenth multiple, checking each other's work. Repeat, choosing different numbers, and different children to start.

ACTIVITY 3
2 pairs

• *Recognising common multiples of two numbers*

Number cards (1 to 9) (PCM 9), counters

The cards are shuffled and placed face down in a pile. One pair chooses two cards and shows them to the second pair. The second pair writes down two numbers which are common multiples of the card numbers. The children check the answers. The second pair collects a counter for each correct multiple. The children continue playing for eight rounds, sharing the roles. Who has the most counters?

ACTIVITY 4
2–3 children

• *Finding the smallest common multiples of three or more numbers*

Number cards (2 to 10) (PCM 9), counters

The children shuffle the cards and turn over three of them. They then find the two smallest common multiples of the three numbers. They discuss the correct answers. The children collect one counter for each correct common multiple. They replace the cards and repeat the activity several times. Who collects the most counters? Extend the activity by asking the children to take four cards.

GROUP ACTIVITY 5
2 pairs

• *Recognising numbers which are common multiples of different numbers*

Multiplication square (1 to 100) (PCM 17), counters

GROUP ACTIVITY 6
Pairs

• *Recognising a number, given a set of its multiples*

Number cards (1 to 50) (PCMs 9 to 12)

Say the multiples

Properties of number
GROUP ACTIVITY 5

2 pairs

Multiplication square (1 to 100), counters

In turn, one pair says a number that appears more than once on the multiplication square, for example 20.

The other pair finds each 20 on the square and says what it is a multiple of: 20 is a multiple of 2, of 4, of 5 and of 10.

Place a counter on each 20 on the square.

Swap roles and continue until there are no more numbers that appear more than once.

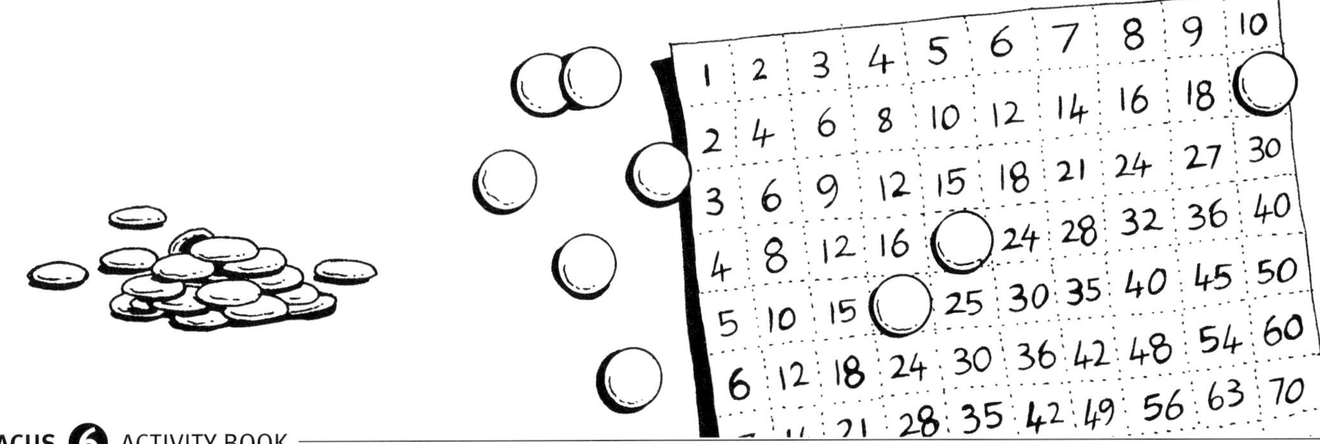

Yes or no

Properties of number
GROUP ACTIVITY 6

Pairs

Number cards (1 to 50)

Shuffle the cards and place them face down in a pile.

One of you choose a number (from 2 to 10) and write it down secretly.

The other person takes the cards from the pile, one at a time, and holds them up.

You say 'yes' if the card number is a multiple of your number, and 'no' if it is not a multiple of your number. Your partner must place the cards in two piles, a 'yes' pile and a 'no' pile, and guess your number.

Continue until your number is guessed.

Count how many cards were needed.

Repeat several times, swapping roles.

Abacus Ginn and Company 2001 Copying permitted for purchasing school only. This material is not copyright free.

N14 Properties of number

ACTIVITY 1
Whole class, in pairs

• *Recognising divisibility by 2, by 4, by 8*

Place-value cards (H, T, even Units) (PCMs 1 to 4)

One child takes one card of each type and creates a 3-digit number, e.g. 346. Each pair has to decide if the number is divisible by 2, by 4, and by 8, recording 'Y' for yes and 'N' for no. For example, (Y, N, N). The children work this out by halving, then halving again. They check each other's answers, scoring one point for each correct set of responses. Repeat several times. Who has the most points?

ACTIVITY 2
2–3 children

• *Recognising divisibility by 3, by 4*

Number cards (20 to 100) (PCMs 10 to 15)

Shuffle the cards and place them in a pile, face down. In turn, the children take a card and decide if the number divides by either 3 or 4. If it does, they keep the card and have another turn. If not, they return the card to the bottom of the pile. Continue until the children have had 12 turns each. Who has the most cards?

ACTIVITY 3
2 pairs

• *Testing for divisibility by 2, 3, 4, 6, 8 and 9*

Place-value cards (H, T, U) (PCMs 1 to 4), counters

Shuffle each set of cards separately and place them in three piles face down. One pair takes a card from each pile to make a 3-digit number. This pair checks the divisibility of the number by 2, 6 and 9. The other pair checks its divisibility by 3, 4 and 8. The pairs take a matching number of counters for each possible divisibility. For example, if the number is divisible by 9, they take nine counters. The children continue until all the cards have been used. Who has the most counters?

ACTIVITY 4
2–3 children

• *Testing for divisibility by the numbers 1 to 10*

Three dice (1 to 6), a calculator, counters

Each child takes turns to throw the three dice to create a 3-digit number and write it down. They then work out the largest number, up to 10, which will divide into the number and take the matching number in counters. For example, if it is divisible by 9, they take nine counters. The children check each other's answers, using a calculator if necessary. They continue until one child has collected 60 counters.

GROUP ACTIVITY 5
3–4 children

• *Testing for divisibility by the numbers 1 to 10*

Three sets of number cards (0 to 9) (PCM 9)

GROUP ACTIVITY 6
3 children

• *Testing for divisibility by the numbers 1 to 10*

A dice (1 to 6), interlocking cubes

Divisibility score

Properties of number
GROUP ACTIVITY 5

3–4 children

Three sets of number cards (0 to 9)

Shuffle the cards and place them face down in a pile.

Each take three cards and make a 3-digit number.

Study your number. Make a list of all the numbers up to 10 that divide equally into it.

Score one point for each.

Think carefully about the 3-digit number you make and try and score as highly as possible.

Check each other's work. Write down each score, shuffle the cards and play again.

3 2 8

328 is divisible by 2, 4, 8

Divisibility testing

Properties of number
GROUP ACTIVITY 6

3 children

A dice (1 to 6), interlocking cubes

Roll the dice three times to make a 3-digit number – the first roll for the hundreds, the second for the tens and the third for the units.

Study the number. Work together to make a list of all the numbers up to 10 that will divide equally into this number.

Build strips of cubes to match each dividing number.

Play again, rolling the dice to make a new 3-digit number.

After ten rounds, look at the strips of cubes.

Which size appears most? Which size appears least?

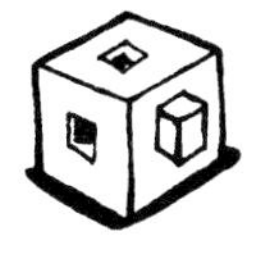
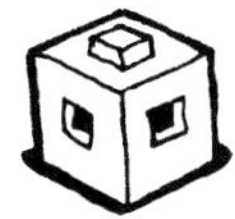
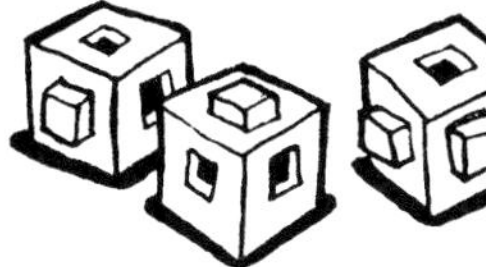

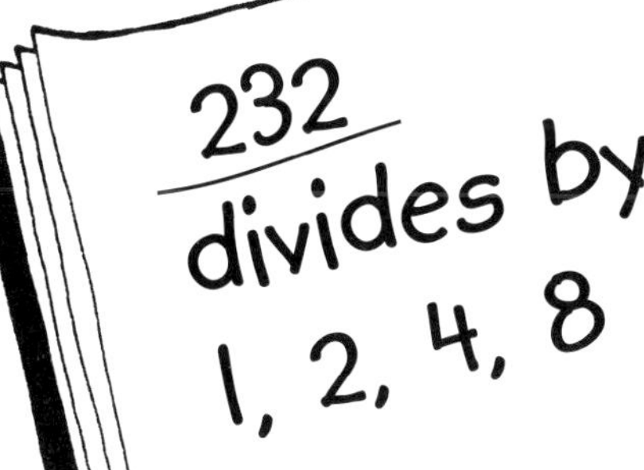

Abacus Ginn and Company 2001 Copying permitted for purchasing school only. This material is not copyright free.

N15 Place-value

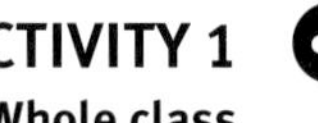
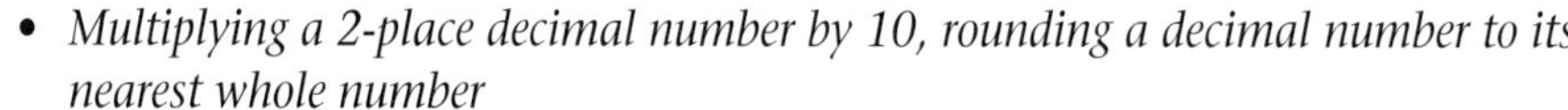

ACTIVITY 1
Whole class

- *Multiplying a 2-place decimal number by 10, rounding a decimal number to its nearest whole number*

Two sets of number cards (1 to 9, and several 0 cards) (PCM 9), Blu-tack
Draw a place-value grid (T, U, t, h) on the board. Ask a child to take a card and stick it in any of the four columns. Let two other children do the same. Place a zero in the empty column. All the children write down the number, e.g. 30·58 and multiply it by 10, i.e. $30{\cdot}58 \times 10 = 305{\cdot}8$. They then round it to the nearest whole number, i.e. 306. Let another child demonstrate the answer by sliding the cards one place to the left. Repeat several times.

ACTIVITY 2
2 pairs

- *Multiplying a 1-place decimal number by 10, and 100*

Place-value cards (U, t) (PCMs 1 to 4)
Shuffle the cards separately and deal five from each set to each pair to make five different decimal numbers, e.g. 3·7, 2·8, 1·6, 9·4, 5·0. Each child multiplies each number by 10 and then by 100. The children check each other's answers. Repeat with a different set of five numbers.

ACTIVITY 3
3–4 children

- *Multiplying a 1-place decimal number by 10, and 100*
- *Estimating and measuring in centimetres using decimal notation*

A collection of objects to measure (pencil, brush, stick), a ruler (marked in cm and mm)
The children estimate the total length of ten and a hundred of each object. Using the ruler, they accurately measure the length of each object in centimetres, e.g. 12·6 cm. They then calculate in metres the length of ten similar objects, when laid end to end, i.e. 1·26 m. Extend the activity by asking the children to calculate the length of 100 similar objects, i.e. 12·6 m. Let the children compare the actual measurements with the estimates.

ACTIVITY 4
3 children

- *Recognising what decimal number has been multiplied by 10 or 100 to make a given number*

Number cards (0 to 9) (PCM 9)
Spread out the cards face up. One child chooses three cards to make a 3-digit number. The second child says what number must be multiplied by 10 to make that number. The third child says what number must be multiplied by 100 to make it. Repeat twice more and then let the children swap roles. Extend the activity by starting with a 2-digit number and a 1-digit number.

GROUP ACTIVITY 5
2–3 children

- *Multiplying decimal numbers by 10, and 100*

Counters

GROUP ACTIVITY 6
3–4 children

- *Multiplying a 2-place decimal number by 10, and 100*

Place-value cards (U, t, h) (PCMs 1 to 4)

Uncover the number

Place-value
GROUP ACTIVITY 5

2–3 children

Counters

Cover each number with a counter.

Take turns to remove a counter.

Multiply the number by 10 and say the answer.

Check each other's answers.

Repeat for multiplying by 100.

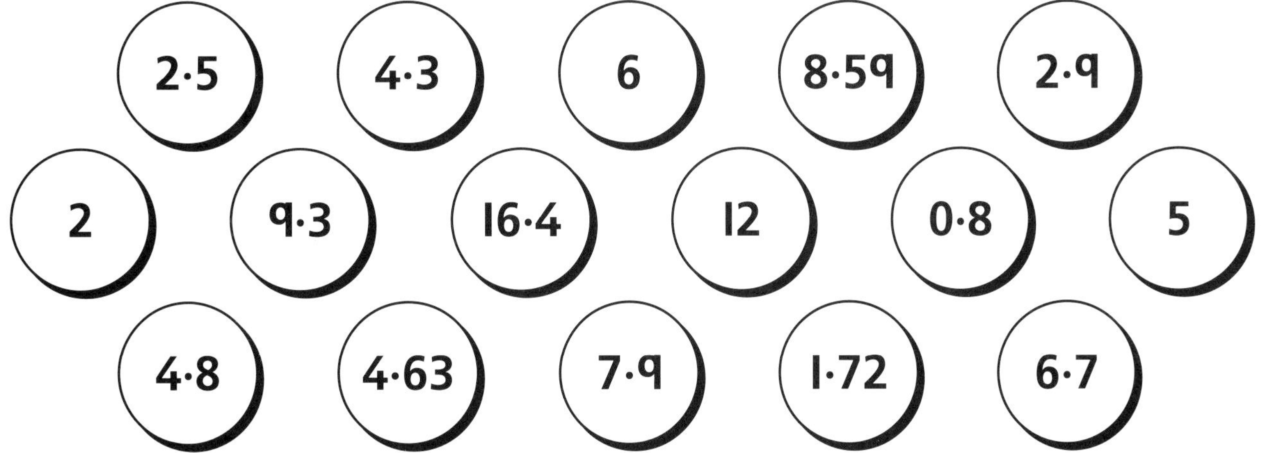

Multiple decimals

Place-value
GROUP ACTIVITY 6

3–4 children

Place-value cards (U, t, h)

Shuffle the cards separately and place them face down in three piles (units, tenths and hundredths).

Each take one card from each pile to make a decimal number.

Repeat eight more times to make nine numbers altogether.

Write them in a table and multiply each by 10 and by 100.

number	x10	x100
3·52	35·2	352
8·16		

Abacus Ginn and Company 2001 Copying permitted for purchasing school only. This material is not copyright free.

 Place-value

ACTIVITY 1
Whole class

- *Dividing a 1-place decimal number by 10, and 100*

Two sets of number cards (1 to 9, and several 0 cards) (PCM 9), Blu-tack

On the board, draw a place-value grid (T, U, t). A child takes a card and sticks it in the tens column. Another child sticks a card in the units column. A third child sticks a card in the tenths column. All the children write down the number, e.g. 25·8 and divide it by 10, i.e. 25·8 ÷ 10 = 2·58. Let another child demonstrate the answer by sliding the cards one place to the right. Extend the activity by dividing the numbers by 100.

ACTIVITY 2
2 pairs

- *Dividing 1- and 2-digit numbers by 10, and 100*

Place-value cards (T, U) (PCMs 1 to 4)

The children shuffle each set of cards separately and create five 2-digit numbers and four 1-digit numbers with them, e.g. 26, 39, 52, 85, 41, 3, 8, 4, 7. Each child divides each number by 10 and then by 100. They check each other's answers. Repeat with a different set of nine numbers.

ACTIVITY 3
3–4 children

- *Dividing a 1-place decimal number by 10, rounding a 1-place decimal number to its nearest whole number*

Place-value board (PCM 16), two sets of number cards (1 to 9, and several 0 cards) (PCM 9)

One child places a card in any of the three columns. Another child does the same. Another child places a zero in the empty column. All the children write the number, e.g. 81·6 and divide it by 10, i.e. 81·6 ÷ 10 = 8·16. They then round it to the nearest whole number, i.e. 8. They check the answer by sliding the cards one place to the right. Repeat several times.

ACTIVITY 4
3 children

- *Recognising what decimal number has been divided by 10 or 100 to make a given number*

Number cards (0 to 9) (PCM 9), a counter

The children spread out the cards face up. One child chooses three cards to make a 1- or 2-place decimal number. The second child says what number must be divided by 10 to make that number. The third child says what number must be divided by 100 to make it. Repeat twice more, then swap roles.

GROUP ACTIVITY 5
3 children

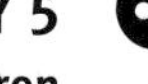

- *Dividing decimal numbers by 10*

Number cards (0 to 9) (PCM 9), a counter, interlocking cubes

GROUP ACTIVITY 6
3–4 children

- *Dividing whole numbers and decimal numbers by 10, and 100*

Counters (3 to 4 colours), a calculator

N16

Find the secret number

Place-value
GROUP ACTIVITY 5

3 children

Number cards (0 to 9), a counter, interlocking cubes

Take turns to take two or three cards and, using the counter as a decimal point, secretly create a number.

Divide the number by 10 and write the answer.

The others look at the answer and work out the secret number. Reveal the number as a check. They collect a cube if they are correct.

Replace the cards and play several times, recording each division together.

Who collects the most cubes?

N16

Grid division

Place-value
GROUP ACTIVITY 6

3–4 children

Counters (three or four colours), a calculator

Copy the grid onto a large piece of paper.

Take turns to choose a number on the grid, and say the result of dividing it by 10.

Write the division.

Use the calculator to check each other's calculation.

If you are correct, place a counter on the grid.

If not, do nothing.

Continue until all the numbers have been covered.

Who has the most counters on the grid?

Repeat the activity, dividing the numbers by 100.

1·3	167	18	7·9	5·4	236
531	43	352	7	4·4	1
9	3·7	9·1	487	35	0·3
0·8	95	26	2·5	130	629
6·3	240	403	0·9	201	86
757	72	5·8	3	8·2	200

Abacus Ginn and Company 2001 Copying permitted for purchasing school only. This material is not copyright free.

N17 Multiplication/division

ACTIVITY 1
Whole class, in pairs

- *Multiplying a 3-digit by a 1-digit number*

Place-value cards (H, T, U) (PCMs 1 to 4), a dice (1 to 6), counters
Shuffle each set of cards separately and place them face down in three piles. A child takes one card from each pile to create a 3-digit number. Another child throws the dice. In pairs, the children multiply the two numbers together and write down their answer. Check the answer together, and give the pairs with a correct answer a counter. Repeat several times to see which pair collects the most counters.

ACTIVITY 2
2–3 children

- *Multiplying an amount up to £1 by a 1-digit number*

Coins (£1, 10p, 1p), number cards (2 to 9) (PCM 9), a cloth bag
Place nine 10p and nine 1p coins in the bag. Shuffle the cards and place them in a pile face down. One child takes a handful of coins, e.g. 45p, another selects a number card, e.g. 3, and all the children write down and complete the multiplication, i.e. 3 × 45p = £1·35. The children check the answer together using the coins, i.e. they collect three lots of 40p and three lots of 5p, and count the total. Repeat several times.

ACTIVITY 3
2 pairs

- *Multiplying a 2-digit by a 1-digit number*

Number cards (2 to 9, 10 to 99) (PCMs 9 to 15), a calculator, interlocking cubes
The cards are shuffled separately and placed face down in two piles. Each pair takes one card from each pile and secretly multiplies the two numbers together. They then exchange number cards and multiply their new numbers together. They then check the results to see if they match, using a calculator if necessary. Each pair collects a cube for every correct answer. Repeat several times. Which pair collects the most counters?

ACTIVITY 4
2–3 children

- *Multiplying a 3-digit number by a 1-digit number, investigating the creation of multiplications which are close to a given product*

Number cards (1 to 9) (PCM 9)
The children select four cards to make a 3-digit number and a 1-digit number. They then multiply them together. They investigate what choice of cards and numbers will give a product which is closest to 1500, 2000, 2500, 3000, ... 7000.

GROUP ACTIVITY 5
Pairs

- *Multiplying a 2-digit number by a 1-digit number*

A calculator, scissors

GROUP ACTIVITY 6
3 children

- *Multiplying a 2-digit number by a 1-digit number, creating multiplications, using three given digits, which are close to a given product*

Number cards (1 to 9) (PCM 9), a calculator

Estimate the order

Multiplication/division
GROUP ACTIVITY 5

Pairs

A calculator, scissors

Cut out the 15 rectangles and arrange them in an estimated order, that is, place the number which you think has the smallest answer on the left, then the next smallest, and so on, with the estimated largest answer on the right.

Next calculate each answer, writing the answer below the multiplication.

Check your estimated order.

How many multiplications were out of place?

4×76	43×7	6×53
9×27	3×85	61×5
52×6	2×91	8×31
87×2	19×9	72×4
5×64	7×48	75×4

Target multiplication

Multiplication/division
GROUP ACTIVITY 6

3 children

Number cards (1 to 9), a calculator

Each make a scoresheet like the one below.

Shuffle the cards and deal three each.

Each create a multiplication of a 2-digit number with a 1-digit number, aiming for an answer as close as possible to the target on your scoresheet.

Multiply the numbers together and write them on the scoresheet. Check each other's answers, using a calculator if necessary.

Find the difference between the answer and the target. The winner is the one whose answer is the closest to the target.

Target	Multiplication	Difference
150		
300		
400		
250		
450		
350		

Abacus Ginn and Company 2001 Copying permitted for purchasing school only. This material is not copyright free.

N18 Multiplication/division

ACTIVITY 1
Whole class, in pairs

• *Multiplying ThHTU × U, using standard written methods*
Place-value cards (Th, H, T, U) (PCMs 1 to 4), a dice (1 to 6), counters
Shuffle the cards separately and place them in four piles, face down. Four children each take a card from one of the piles to create a 4-digit number, e.g. 3174. Another child rolls the dice to find the multiplier, e.g. 4. (If 1 turns up, they roll it again.) Each pair multiplies the two numbers together, i.e. 4 × 3174, splitting it into four separate multiplications, then adding the answers. Discuss the correct answer together. Pairs with a correct answer collect a counter. Repeat several times. Who collects the most counters?

ACTIVITY 2
3–4 children

• *Multiplying ThHTU × U, using standard written methods*
Number cards (0 to 9) (PCM 9), a calculator, counters
Play 'Multiplying by 2'. The children shuffle the cards and turn over four to make a 4-digit number. They multiply the number by 2, then check each other's answers, using a calculator if necessary. They take a counter if they are correct. They put the cards back into the pack, turn over four more and play 'Multiplying by 3'. They continue up to multiplying by 9. Who has the most counters?

ACTIVITY 3
3–4 children

• *Estimating a multiplication, HTU × TU*
Number cards (1 to 9) (PCM 9), a calculator
Spread out the cards face down. The children turn over five cards to make a 2-digit number and a 3-digit number. Each child rounds both numbers to their nearest 10, then multiplies the two to estimate the answer. They check each other's rounding and estimates. One child multiplies the two numbers using the calculator. How close are the estimates? Repeat several times with the children sharing the roles.

ACTIVITY 4
2 pairs

• *Estimating the result of multiplying HTU × TU, multiplying HTU × TU, using standard written methods*
A calculator, counters
Each pair writes a 2-digit number and a 3-digit number (excluding multiples of 10), which they estimate will make 3000 when multiplied together. They swap and multiply each other's numbers and then swap back numbers for checking. They can use a calculator, if necessary, as a further check. The pair whose numbers are nearest to 3000 collects a counter. Repeat with a target number of 4000. Continue, increasing the target by 1000 each time.

GROUP ACTIVITY 5
3–4 children

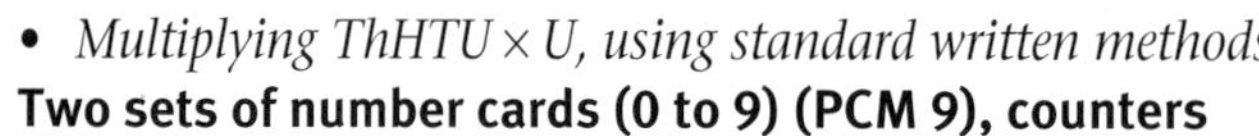

• *Multiplying ThHTU × U, using standard written methods*
Two sets of number cards (0 to 9) (PCM 9), counters

GROUP ACTIVITY 6
3–4 children

• *Multiplying HTU × TU, using informal written methods*
Number cards (1 to 9) (PCM 9)

Five multiplications

N18

Multiplication/division
GROUP ACTIVITY 5

3–4 children

Two sets of number cards (0 to 9), counters

Arrange the number cards to make five 4-digit numbers.
Choose a 1-digit multiplying number, for example, 4.

Each multiply the card numbers, in turn, by your chosen number.
Check each other's answers.

Collect a counter for each correct answer.

Repeat five times, choosing different multiplying numbers and making different 4-digit numbers each time.

Who collects the most counters?

Three by two

N18

Multiplication/division
GROUP ACTIVITY 6

3–4 children

Number cards (1 to 9)

Choose a number between 12 and 20, for example 16.

Deal out three cards each to make a 3-digit number. Each write the multiplication of your second number by your first.

Each draw a rectangle to match the multiplication. Split the number into hundreds, tens and units. Write the areas inside each block, and then find the total.

Check each other's answers.

Repeat several times.

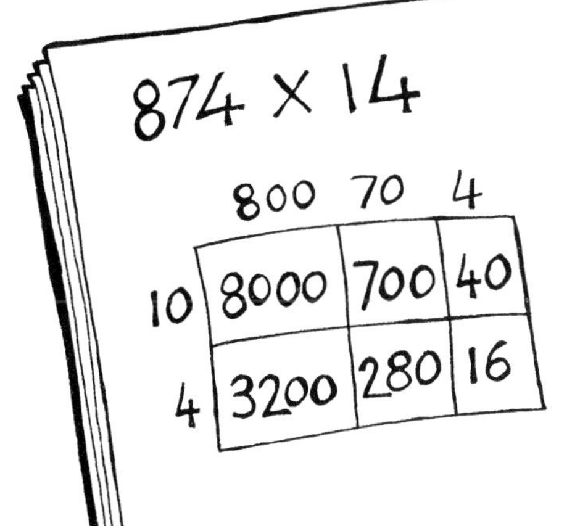

Abacus Ginn and Company 2001 Copying permitted for purchasing school only. This material is not copyright free.

N19 Multiplication/division

ACTIVITY 1
Whole class, in pairs

- *Constructing two division facts from one multiplication fact*

Number cards (1 to 9, 10 to 99) (PCMs 9 to 15), Blu-tack

Shuffle the cards separately and place them face down in two piles. Let two children take one card from each set, e.g. 36 and 7, and stick them on the board. The children use these cards to create two division facts, first by multiplying the numbers together, e.g. $7 \times 36 = 252$, which produces $252 \div 7 = 36$ and $252 \div 36 = 7$. Check the results together. Give the children one point for each correct statement. Repeat several times. Which pair has the most points?

ACTIVITY 2
3–4 children

- *Constructing multiplication facts and their associated division facts*

Number cards (1 to 9) (PCM 9)

The children draw a 4×4 grid as outline for a multiplication table They use six of the card numbers as headings, three as column headings and three as row headings. They complete the multiplication table with nine products. They then write two division facts for each product.

ACTIVITY 3
2 pairs

- *Deducing a multiplication fact given the answer, constructing a division fact from a known multiplication fact*

Place-value cards (U, t) (PCMs 1 to 4), two dice (1 to 6), counters

Each pair secretly selects one of each type of card to create a decimal number, e.g. 4·7, then throws the dice to find a multiplier, e.g. 5. They find the product, i.e. 23·5 and give it to the other pair. This pair then has to deduce a matching division fact from the product. If their answer is correct, they receive a counter. Repeat several times. Which pair has the most counters?

ACTIVITY 4
Pairs

- *Constructing multiplication facts and their associated division facts*

Spreadsheet software

Prepare a spreadsheet file similar to the one shown. Use the formula **=A2*C2** in Column E, then fill in numbers in columns A and C to set up some multiplication facts. Use the 'If logical' function **=IF(K2>0,IF(K2=G2/I2,"Well done","Try again")," ")** in Column L. The children use the multiplication facts in Columns A to E to fill in the numbers in columns G, I and K to show a matching division fact. They check Column L to see if their answers are correct.

	A	B	C	D	E	F	G	H	I	J	K	L
1		multiplied by		equals				divided by		equals		
2	69	×	5	=	345		345	÷	5	=	69	Well done
3	175	×	3	=	525		175	÷	3	=	525	Try again
4	131	×	4	=	524			÷		=		
5	103	×	7	=	721			÷		=		

GROUP ACTIVITY 5
3–4 children

- *Constructing multiplication facts and their associated division facts*

GROUP ACTIVITY 6
3 children

- *Multiplying a 3-digit number by a 1-digit number, constructing division facts from a known multiplication fact*

Seven multiplications

3–4 children

13·5	93	4	8·4	39	9·5	5
2	42	5	7·8	27	8	48
5	15·5	58	14·5	6	9·6	76

Find seven different multiplication facts using sets of three of the numbers above.

Write two matching divisions for each.

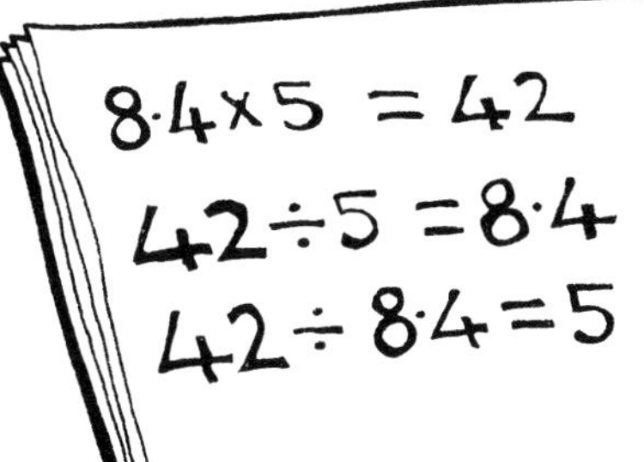

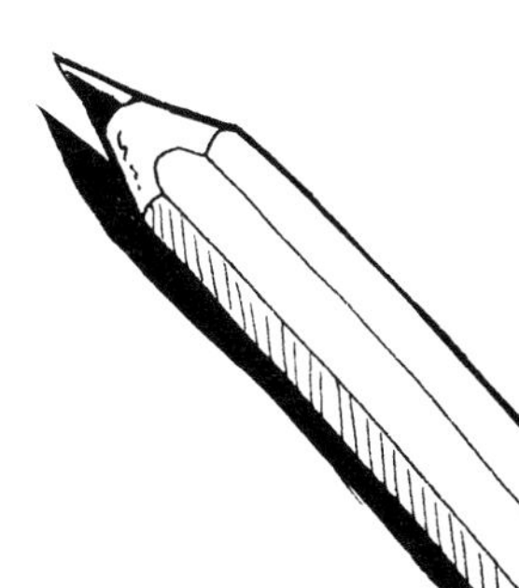

Division facts

3 children

Copy and complete the multiplication table, then use the results to complete the division facts.

882 ÷ 7 =

705 ÷ 3 =

630 ÷ 126 =

1645 ÷ 235 =

378 ÷ 3 =

504 ÷ 126 =

1570 ÷ 5 =

1256 ÷ 4 =

2198 ÷ 314 =

942 ÷ 3 =

1175 ÷ 5 =

940 ÷ 235 =

×	4	3	5	7
126				
314				
235				
186				

Abacus Ginn and Company 2001 Copying permitted for purchasing school only. This material is not copyright free.

N20 Multiplication/division

ACTIVITY 1
Whole class, in pairs

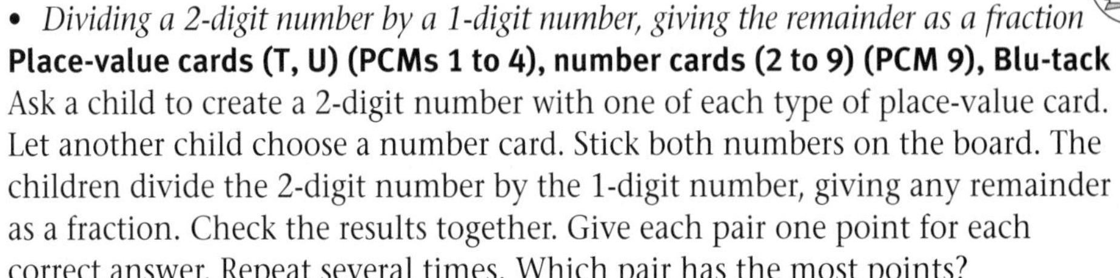

- *Dividing a 2-digit number by a 1-digit number, giving the remainder as a fraction*

Place-value cards (T, U) (PCMs 1 to 4), number cards (2 to 9) (PCM 9), Blu-tack

Ask a child to create a 2-digit number with one of each type of place-value card. Let another child choose a number card. Stick both numbers on the board. The children divide the 2-digit number by the 1-digit number, giving any remainder as a fraction. Check the results together. Give each pair one point for each correct answer. Repeat several times. Which pair has the most points?

ACTIVITY 2
3 children

- *Dividing a 2-digit by a 1-digit number, giving the remainder as a fraction*

Multiplication square (1 to 100) (PCM 17)

Children take turns to say a particular multiple, e.g. ×5. The other children say any number less than the tenth multiple, i.e. 50, which does not appear in the fifth row of the multiplication square, e.g. 37. The first child divides this number by 5, expressing the remainder as a fraction, i.e. $7\frac{2}{5}$. They can use the numbers in the fifth row of the multiplication square to help them. Repeat several times.

ACTIVITY 3
2 pairs

- *Dividing a 2-digit number by each number from 2 to 12, giving the remainder as a fraction, dividing by 10, giving the remainder as a decimal*

Number cards (50 to 99) (PCMs 12 to 15)

The cards are shuffled and placed in a pile, face down. One child reveals the top card. All the children divide the number by 2, then 3, then 4, ... up to 12, expressing each remainder as a fraction. When dividing by 10, they should also express the remainder as a decimal. The children check each other's answers, scoring one point for each correct answer. Repeat several times. Who collects the most points?

ACTIVITY 4
2–4 children

- *Dividing, giving the remainder as a decimal, calculating an average (mean)*

Two to four reading books

Each child opens a book at a random page and counts the number of words on ten lines. They find the average (mean) number of words per line by dividing one by the other, and expressing the remainder as decimal. The children exchange books and repeat the activity for each book. They then compare their results.

GROUP ACTIVITY 5
2 pairs

- *Dividing a 2-digit by a 1-digit number, giving the remainder as a fraction*

A dice (1 to 6), counters (two colours)

GROUP ACTIVITY 6
2–3 children

- *Creating a division which produces a given mixed number*

Counters

Three in a line

2 pairs

A dice (1 to 6), counters (two colours)

Make a large copy of the grid.

One pair chooses a number on the grid. The other pair rolls the dice and divides the grid number by the dice number, giving the remainder as a fraction.

Both pairs do the calculation to check.

If it is correct, the second pair places one of their counters over the chosen number on the grid. If it is incorrect, then the first pair covers it with one of their counters.

Continue until one pair has three counters in a straight line.

58	22	42	13	51
11	31	46	19	44
27	53	23	35	39
41	37	49	17	43
15	47	36	45	38

Guess the division

2–3 children

Counters

Cover each circle below with a counter. On your turn, the other players choose a counter for you to remove.

Write a division which has the revealed number as an answer. For example, if you reveal $3\frac{5}{6}$, then you write 23 ÷ 6.

If you are correct, keep the counter. If not, replace it.

When all the counters have been removed, who has the most?

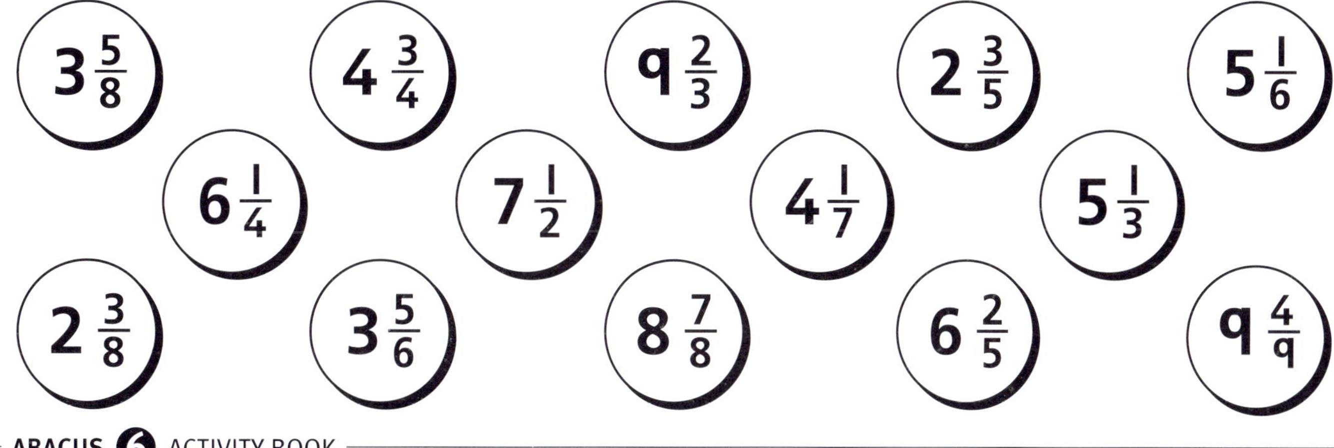

Abacus Ginn and Company 2001 Copying permitted for purchasing school only. This material is not copyright free.

N21 Fractions/decimals

ACTIVITY 1
Whole class, in groups of 3

- *Working out fractions of amounts of money*

Coins (1p, 10p, £1), post-it notes

One child writes a mixed fraction on a post-it note, e.g. $2\frac{1}{4}$. The other two children work out the matching number of pence and pounds shown by that fraction and select the appropriate coins. They write the amount on the reverse of the note, e.g. £2·25p. They check that they all agree on this amount. Repeat ten times with the children sharing the roles. The children then arrange the notes with the fractions showing. Each child chooses a note and says what amount is written on the reverse. Are they correct?

ACTIVITY 2
3–4 children

- *Working out fractions of time*

Strips of paper

Each child uses a strip of paper to make a train ticket, writing the journey time as a mixed fraction, e.g. Train: $4\frac{2}{3}$ hours. The children swap tickets, and look at the mixed fraction. They write the time in hours and minutes on the reverse of the ticket, i.e. 4 hours and 40 minutes. They check each other's tickets and repeat the process. When they have made at least 12 train tickets, they spread them out with the mixed fractions showing. They take turns to choose a ticket, say the time in hours and minutes and turn the ticket over to see if they are correct.

ACTIVITY 3
3 children

- *Estimating fractions of amounts*

A small straight-sided mug, dried beans, post-it notes

Using the mug, one child secretly measures out an amount of dried beans, e.g. two and a half mugfuls. The other two children estimate how many mugfuls of beans there are. They write their estimate as a mixed fraction on a post-it note. They then measure the beans using the same mug and write the measurement below their guess. How close were they? Repeat ten times, with the children taking different roles.

GROUP ACTIVITY 4
3–4 children

- *Converting improper fractions into mixed numbers*

A dice (1 to 6)

GROUP ACTIVITY 5
3–4 children

- *Converting improper fractions into mixed numbers*
- *Working out fractions of whole numbers*

Number cards (12 to 20) (PCM 10), a dice (1 to 6)

Five star

Fractions/decimals
GROUP ACTIVITY 4

3–4 children

A dice (1 to 6)

Take turns to play. Choose one of the star numbers shown. Roll the dice. This is the denominator.

Write the fraction. For example, you choose 25, roll a 4 and write $\frac{25}{4}$.

Each turn the fraction into a mixed fraction, for example, $6\frac{1}{4}$.

Compare the fractions. If your fraction is closest to 5, you win that round and score five points.

Keep playing until one of you scores 30 points.

Biscuit score

Fractions/decimals
GROUP ACTIVITY 5

3–4 children

Number cards (12 to 20), a dice (1 to 6)

Shuffle the cards and place them in a pile face down.

Take turns to play. Take a card, roll the dice and create a fraction using the card as the top number and the dice number as the denominator. For example, you take card 13, roll a 3 and make $\frac{13}{3}$.

Turn the fraction into a mixed fraction, for example, $4\frac{1}{3}$.

Choose a biscuit packet, one with either 10 or 12 biscuits, and work out how many biscuits you can have in $4\frac{1}{3}$ packets. You may not have half or part biscuits so choose carefully. Score the number of biscuits you can take. Continue playing, taking it in turns. Who has had most biscuits after ten turns each?

Abacus Ginn and Company 2001 Copying permitted for purchasing school only. This material is not copyright free.

N22 Fractions/decimals

ACTIVITY 1
Whole class

- *Adding to a 3-place decimal number*

Place-value cards (U, t, h, th) (PCMs 3 to 7), Blu-tack

Shuffle each set of cards and place them face down in four piles. Ask four children to take one card each from a different pile. The child with the units card sticks it on the board, then the child with the tenths card does the same, and so on, until they have created a 3-place decimal number. The children read the number together. On the board one child writes the number that is one thousandth more than that number, another writes the number that is one hundredth more and another writes the number that is one tenth more. The children check each other's numbers. Repeat ten times with different children.

ACTIVITY 2
4 children

- *Creating and ordering 3-place decimal numbers*

Place-value cards (U, t, h, th) (PCMs 3 to 7), post-it notes

Shuffle each set of cards and place them face down in four piles. Each child takes a card from a different pile. The child with the units card places their card face up on the table, then the child with the tenths does the same, and so on, until they have a 3-place decimal number. The children read the number together and one of them writes it on a post-it note. They repeat this until they have made ten different numbers. They then arrange all the notes in order from smallest to largest. One child points at two of the numbers and the others have to say a number which comes between them. One of them writes it on a post-it note and positions it in the line. Repeat five more times, with the children sharing roles.

ACTIVITY 3
4 children

- *Comparing 3-place decimal numbers*

Place-value cards (U, t, h, th) (PCMs 3 to 7), cubes

Shuffle each set of cards and place them face down in four piles. Each child takes a card from each pile and creates a 3-place decimal number. They write down their number and then compare them. The child with the number closest to 4·999 takes a cube. They replace the cards in their appropriate piles and continue until one child has five cubes.

ACTIVITY 4
3 children

- *Creating and ordering 2-place decimal numbers*

A dice (1 to 6), strips of paper, cubes

The children take turns to roll the dice three times. They each use the numbers rolled to create a 2-place decimal number. For example, by rolling a 3, a 2 and a 3, they could make 3·23. They then write these numbers in order, from smallest to largest, along a strip of paper. The child whose number is in the middle takes a cube. Repeat at least ten times. Who has the most cubes?

ACTIVITY 5
3–4 children

- *Comparing and ordering 2-place decimal numbers*

Game 5: 'Out of Order!', cubes, a dice (1 to 6), counters

(See instructions on the card.)

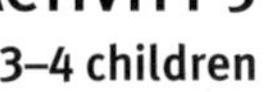

GROUP ACTIVITY 6
2 pairs

- *Writing measurements to three decimal places*
- *Estimating measurements*

String, a 2 m ruler, scissors, post-it notes

GROUP ACTIVITY 7
3 children

- *Creating 3-place decimal numbers*

Three sets of number cards (0 to 9) (PCM 9), cubes

Decimal strings

Fractions/decimals
GROUP ACTIVITY 6

2 pairs

String, a 2 m ruler, scissors, post-it notes

Each pair cuts a length of string and measures it very carefully in metres and centimetres.

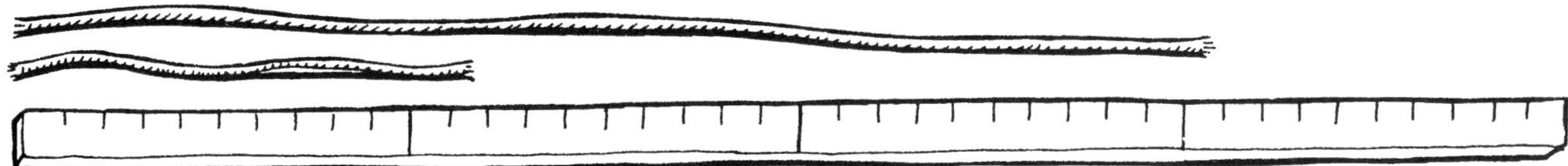

Write the length of the string in metres to three decimal places on a post-it note. For example, if your string is 56 and a half centimetres long, you write 0·565.

Swap strings with the other pair but do not show them your post-it note.

Write an estimate of the length of this second piece of string. Show the other pair your estimate and compare it with the actual length. Which pair was closest? Repeat five times.

Don't move!

Fractions/decimals
GROUP ACTIVITY 7

3 children

Three sets of number cards (0 to 9), cubes

Shuffle the cards and place them in a pile face down.

Each draw a grid like this:

Units	Tenths	Hundredths	Thousandths
•			

Take turns to take a card and place it in one of the spaces on your grid. Once you have placed it, you must not move it.

After four rounds, compare your numbers. The player whose number is closest to 5 takes a cube.

Replace the cards at the bottom of the pile and continue playing. Who has the most cubes after ten games?

Abacus Ginn and Company 2001 Copying permitted for purchasing school only. This material is not copyright free.

N23 Addition/subtraction

ACTIVITY 1
Whole class, in pairs

• *Adding 4-digit numbers*

Place-value cards (Th, H, T, U) (PCMs 1 to 4), a calculator

Shuffle each set of cards and place them face down in four piles. Ask four children to take one card from each pile to create a 4-digit number. They each write their number on the board and all the children copy the four numbers. The children add the numbers together and compare their answers. When they agree a total, they use a calculator to check it. Repeat six times with different children choosing the cards.

ACTIVITY 2
3 children

• *Calculating the numerical value of letters in a vertical addition*

Write this sum on the board as a vertical addition:

```
   W I N D
 + R A I N
 S T O R M
```

Ask the children to copy it. Tell them that each letter is one digit and that no digit has more than one letter. Ask them to work out what digit each letter represents to make the addition work. There are 8 possible solutions:
8579 + 3457 = 12 036, 8769 + 4276 = 13 045, 6428 + 7542 = 13 970, 9328 + 6432 = 15 760, 9346 + 8234 = 17 580, 3526 + 7452 = 10 978, 8264 + 9326 = 17 590, 3254 + 7625 = 10 879

ACTIVITY 3
2 pairs

• *Estimating and calculating the addition of four 3-digit numbers*

Interlocking cubes

Each child writes down a 3-digit number. They look at what the others have written. Each pair estimates the total of all the numbers and writes it down. They all write down each other's numbers and add them. Each pair works together to agree an answer. The pairs then compare each other's answers. Do they agree? Each pair then compares each estimate with the correct answer. The pair whose estimate was closest takes a cube. Repeat ten times. Which pair has the most cubes?

ACTIVITY 4
Pairs

• *Adding three 3-digit numbers to make 1000*

Spreadsheet software

Prepare a spreadsheet file similar to the one shown. Use the 'If logical' function **=IF(E2>0,ABS(1000–(A2+C2+E2))," ")** in Column H. In Column E, the children estimate the number that must be added to those in Columns A and C to make 1000. They check Column H to see how close their estimates were.

	A	B	C	D	E	F	G	H	I
1	324	+	356	+	300		You are	20	away from 1000
2	543	+	213	+	230		You are	14	away from 1000
3	126	+	691	+			You are		away from 1000
4	337	+	486	+			You are		away from 1000
5									

GROUP ACTIVITY 5
3 children

• *Estimating and calculating the addition of three 3- or 4-digit numbers*

Counters

GROUP ACTIVITY 6
3–4 children

• *Adding three 3-digit numbers to make 1000*

Three to four sets of number cards (1 to 9) (PCM 9)

Three-spot addition

Addition/subtraction
GROUP ACTIVITY 5

3 children

Counters

Cover each circle with a counter.

Each remove a counter and write down the three numbers. Estimate their total and write it down.

Each add the three numbers and compare your answers. If your estimate is the closest to the answer, you keep a counter.

Replace all but one of the counters. Keep playing until all the counters have gone.

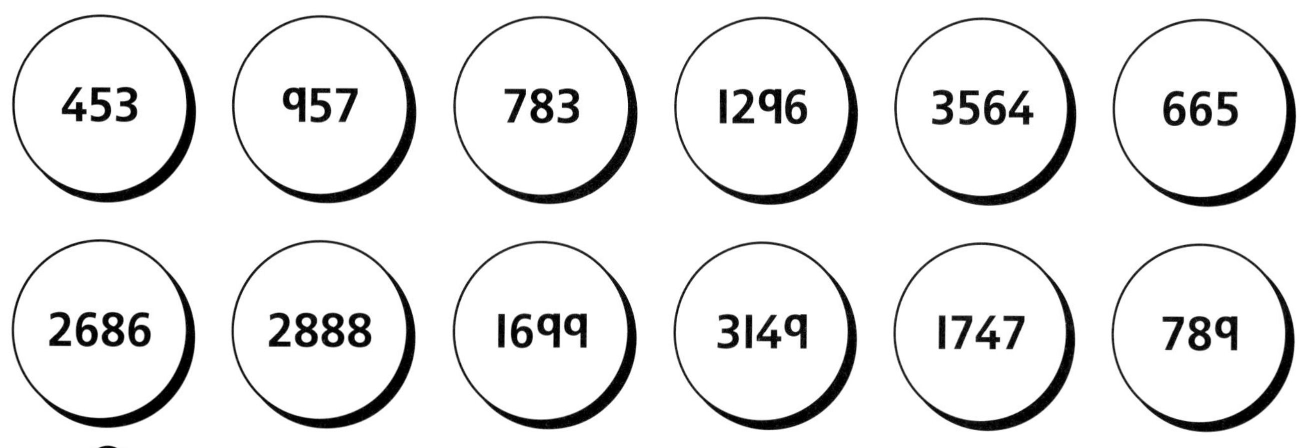

What's your guess?

Addition/subtraction
GROUP ACTIVITY 6

3–4 children

Three to four sets of number cards (1 to 9)

Use the number cards to make three 3-digit numbers.
Try to make numbers which, when added, have a total close to 1000.

Add each set of three numbers and check each other's totals. If your total is closest to 1000 you score 10 points. Keep playing until someone has 100 points.

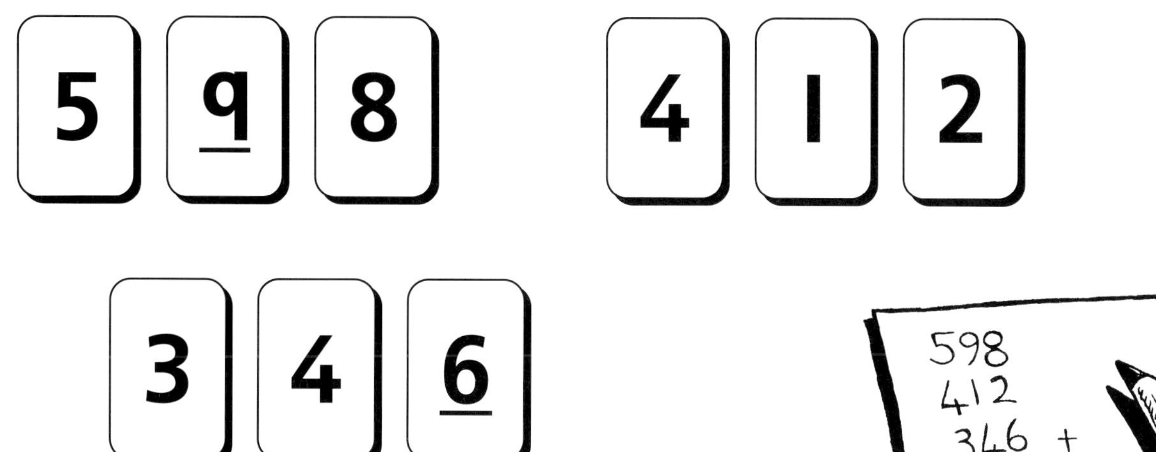

Abacus Ginn and Company 2001 Copying permitted for purchasing school only. This material is not copyright free.

N24 Addition/subtraction

ACTIVITY 1
Whole class, in pairs

- *Adding 2-place decimal numbers*

Place-value cards (U, t, h) (PCMs 3 to 6), a calculator

Shuffle each set of cards and place them face down in three piles. Ask three children to take one card from each pile to create a 2-place decimal number. They each write their number on the board and all the children copy the three numbers. The children add the numbers together and compare their answers. When they agree a total, they use a calculator to check it. Repeat the activity six times with different children choosing the cards.

ACTIVITY 2
3 children

- *Adding 2-place decimal numbers*

Place-value cards (U, t, h) (PCMs 3 to 6), counters

Shuffle each set of cards and place them face down in three piles. Each child chooses either units, tenths or hundredths. They each take one card from their chosen pile to create a decimal number. They repeat the process, estimate the total of the two numbers and then add them together. Each child then collects counters to match the value of their chosen digit. For example, if the total is 6·48, the child who chose hundredths collects eight counters. Play ten rounds. Who has the most counters?

ACTIVITY 3
2 pairs

- *Estimating and calculating the addition of 2-place decimal numbers*

Place-value cards (U, t, h) (PCMs 3 to 6), counters

Shuffle the cards separately and place them face down in three piles. Using two cards from each pile, each pair creates two numbers, estimates the total and writes it down. They then work out the total. They give the other pair their numbers, without the total, and add each other's numbers. They then compare totals. Any pair whose estimate is within 0·5 of the actual total collects a counter. The children replace the cards and play until one pair has eight counters.

ACTIVITY 4
3 children

- *Finding different ways of obtaining a given amount of money*

Coins (one £2, one £1, three 50p, one 20p, one 10p, two 5p, four 2p, two 1p)

The children divide the coins into three piles to make three amounts which add up to £5. Explain that there must be at least three coins in each pile. They write down the addition. How many variations can they find?

GROUP ACTIVITY 5
3 children

- *Adding 2-place decimal numbers*

Two dice (1 to 6), counters

GROUP ACTIVITY 6
Pairs

- *Measuring an irregular 4-sided shape*
- *Adding four 1- or 2-place decimal numbers*

A 25 cm piece of string, a ruler

Two dice addition

Addition/subtraction
GROUP ACTIVITY 5

3 children

Two dice (1 to 6), counters

Roll two dice. The numbers rolled show which decimal numbers you must add together.

Write down the numbers, then estimate the total. Check each other's estimates.

Each add the two numbers, then check each other's additions.

Collect a counter if you are correct. Continue until all the counters have gone. Who has the most counters?

Dice numbers					
1	2	3	4	5	6
3·67	4·15	5·92	3·79	4·48	6·93

String addition

Addition/subtraction
GROUP ACTIVITY 6

Pairs

A 25 cm piece of string, a ruler

Make the string slightly damp and use it to make an irregular 4-sided shape.

Measure each side as accurately as you can.

Write down the length of each side.

Add the four lengths together, writing the addition.

How close is your addition to 25 cm?

Repeat, making another 4-sided shape.

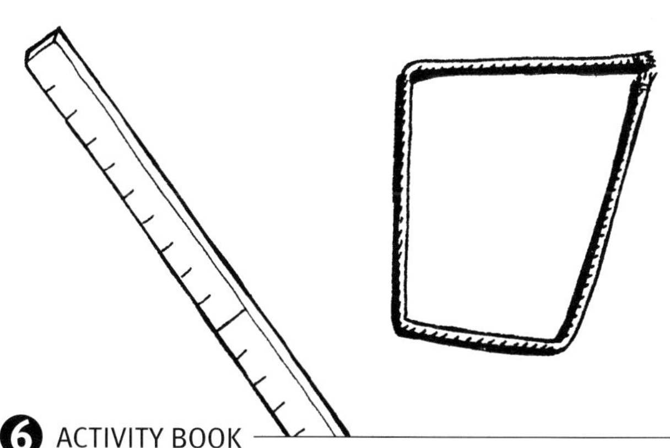

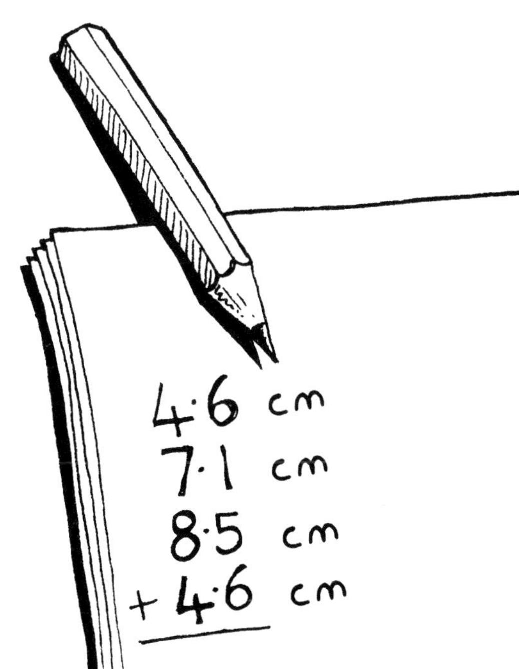

Abacus Ginn and Company 2001 Copying permitted for purchasing school only. This material is not copyright free.

N25 Addition/subtraction

ACTIVITY 1
Whole class, in pairs

- *Subtracting large numbers*

Each child writes down their first and last names. They then write the letters in both names in alphabetical order. They assign a number between 1 and 9 (inclusive) to each letter. (If they have more than ten letters, they start again at 1.) The children write their name again, this time writing numbers instead of letters. They each write down both 'name numbers' and subtract the smaller number from the larger. They check each other's work. Repeat at least six times, suggesting that the children use different names such as Matilda Wormwood or James Bond. Let some of the children demonstrate their 'name numbers' to the rest of the class.

ACTIVITY 2
Pairs

- *Subtracting 4-digit numbers*

Three sets of number cards (0 to 9) (PCM 9), counters

The cards are shuffled and placed face down in a pile. Each child secretly writes down a 4-digit number. Then they take four cards each and make another 4-digit number with them. Each child writes down both numbers and finds the difference, subtracting the smaller number from the larger. They compare answers and agree the correct difference. Then they each subtract the difference from their secret number, writing the subtraction if necessary. The child with the smaller answer, i.e. the one whose secret number is closest to the difference, is the winner and takes a counter. The children replace the cards and play again until one of them has six counters.

ACTIVITY 3
Pairs

- *Subtracting 3- or 4-digit numbers*

A history book including dates

The children look at the history book together and each chooses an event. The events must be at least 100 years apart, for example, the Great Fire of London 1666, the Beatles' first LP 1963. They each write down the date of the event. They then subtract the earlier date from the later one and work out the difference between the two. They must agree the correct answer and then write the two dates and the difference on top of a sheet of paper. They repeat this six times, choosing different dates each time.

ACTIVITY 4
3 children

- *Subtracting 3-digit numbers*

Place-value cards (H, T, U) (PCMs 1 to 4), counters

The cards are shuffled and placed in three piles face down. Each child chooses either hundreds, tens or units. They each take a card from their pile and combine them to make a 3-digit number. They do it again to make a second number. They each write down both numbers and subtract the smaller number from the larger. They compare their answers and agree a correct subtraction. Each then takes the same number of counters as the value of their chosen digit in the answer. Repeat until all the cards have been used. Who has collected most counters?

GROUP ACTIVITY 5
3 children

- *Subtracting 5-digit numbers*
- *Finding a digital root*

A dice (1 to 6), counters

GROUP ACTIVITY 6
2 pairs

- *Subtracting 4-digit numbers*

Place-value cards (Th, H, T, U) (PCMs 1 to 4)

Reducing the score

Addition/subtraction
GROUP ACTIVITY 5

3 children

A dice (1 to 6), counters

Draw a 5 by 3 grid.

Take turns to play. Roll the dice five times, writing each number in a space in the second row of the grid.

Re-arrange the digits to make the largest possible number and write this along the top row.

Subtract the second number from the first.

Add the digits of the answer and add the digits of that answer until you are left with a 1-digit number. That number is your score.

Keep playing until one of you scores over 60.

6	5	2	2	1
6	1	2	5	2
0	3	9	6	9

65221
− 61252
3969 → 3 + 9 + 6 + 9
= 27 →
7 + 2 = 9
Score 9

Make 4999

Addition/subtraction
GROUP ACTIVITY 6

2 pairs

Place-value cards (Th, H, T, U)

Shuffle the cards separately and place them in four piles face down.

Each pair takes two cards from each pile and arranges them to make two 4-digit numbers.

Subtract the smaller number from the larger one. Check each other's subtraction.

The pair with the answer closest to 4999 scores ten points.

Keep playing until one pair scores over 60.

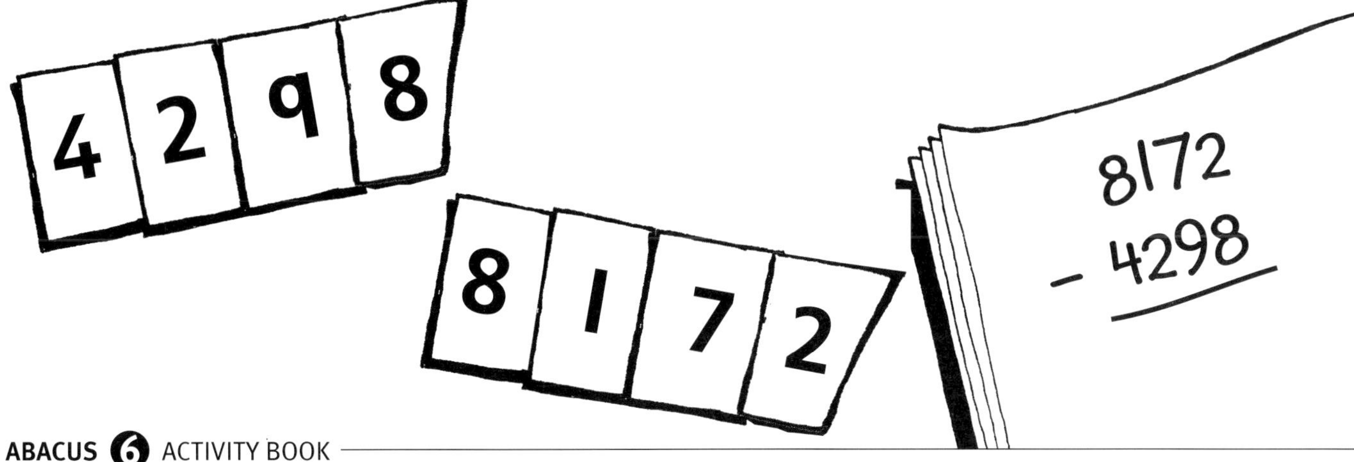

Abacus Ginn and Company 2001 Copying permitted for purchasing school only. This material is not copyright free.

N26 Addition/subtraction

ACTIVITY 1
Whole class, in pairs

- *Subtracting 2-place decimal numbers*

Place-value cards (U, t, h) (PCMs 3 to 6), a calculator, counters

Shuffle each set of cards and place them in three piles face down. Ask two children to choose one card from each set to create two decimal numbers which all the children write down. They then write down the estimate of the difference between the two numbers. Each child writes the subtraction, taking the smaller number from the larger. They compare their answers, then use a calculator to check. Each child then compares their estimate with the answer and those within 0·5 of the actual difference may take a counter. Repeat at least ten times with different children choosing the cards. Who finishes with the most counters?

ACTIVITY 2
3 children

- *Subtracting 2-place decimal numbers*
- *Calculating amounts of money equivalent to decimal numbers*

Three sets of number cards (0 to 9) (PCM 9), coins (£1, 10p, 1p)

The cards are shuffled and placed face down in a pile. The children take six cards to make two 2-place decimal numbers. They use the coins to make a matching amount of money, e.g. one of the numbers is 3·62 and they match this to three £1 coins, six 10p coins and two 1p coins. They write down the numbers they created and subtract the smaller one from the larger. They then check this using coins. How many do they have to add to the smaller amount to make the larger amount? They replace the cards and coins and repeat ten times.

ACTIVITY 3
3 children

- *Subtracting amounts of money*

Coins (£1, 10p, 1p), a cloth bag, counters

Place the coins in the bag. Each child secretly writes down an amount of money. One child takes a handful of coins from the bag, counts the money and writes down the amount. Each child then does a vertical written subtraction to work out the difference between their secret amount and the amount on the table. They check each other's subtractions and compare answers. The child whose difference is closest to £1·99 takes a counter. They repeat at least ten times.

ACTIVITY 4
Pairs

- *Subtracting 3-place decimal numbers*

Three sets of number cards (0 to 9) (PCM 9), counters

Shuffle the cards and place them face down in a pile. Each child secretly writes down a 3-place decimal number then takes four cards to make another. They write down the number they created with the cards and find the difference between it and their secret number, subtracting the smaller from the larger. They check each other's answers. They then each subtract their difference from their secret number. The child with the smaller answer, i.e. the one whose secret number is closest to the difference, is the winner and takes a counter. They replace the cards and play again until one of them has six counters.

ACTIVITY 5
3–4 children

- *Subtracting 2-place decimal numbers*

Game 6: 'Take it away', counters, a dice (1 to 6)

(See instructions on the card.)

GROUP ACTIVITY 6
3–4 children

- *Subtracting 2-place decimal numbers*

Number cards (0 to 8) (PCM 9), cubes

GROUP ACTIVITY 7
3–4 children

- *Subtracting 2-place decimal numbers*
- *Finding a reduced number*

Number cards (0 to 8) (PCM 9), a dice (1 to 6), cubes

N26

Round to score

3–4 children

Addition/subtraction
GROUP ACTIVITY 6

Number cards (0 to 8), interlocking cubes

Make a large copy of the grid shown. Write 9 in the first box.

Take turns to play. Deal five cards into the empty boxes on your grid.

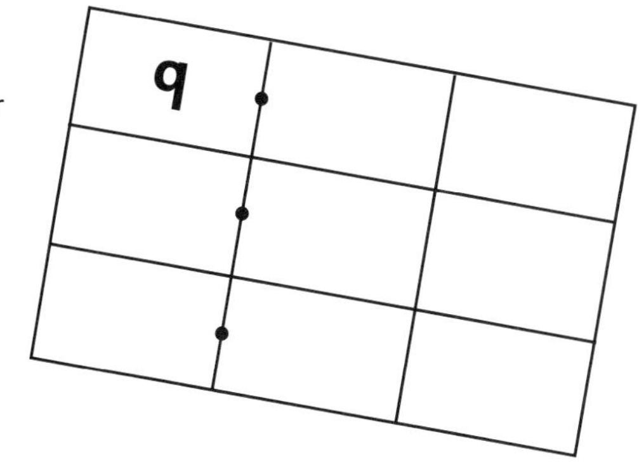

Subtract the number in the second row from the number in the first row. Write the answer in the third row.

The other children work out the subtraction to check your answer.

Round the answer to the nearest whole number. Collect cubes to match this rounded answer.

Play seven times. Who collects the most cubes?

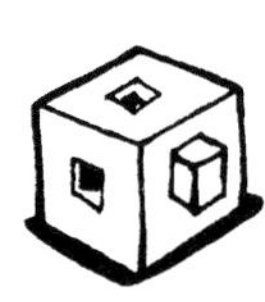   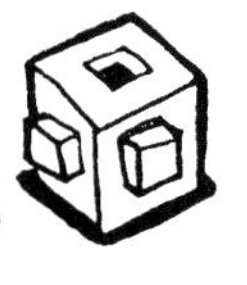 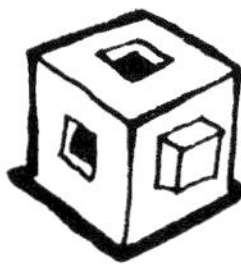

N26

Decimal subtractions

3–4 children

Addition/subtraction
GROUP ACTIVITY 7

Number cards (0 to 8), a dice (1 to 6), interlocking cubes

Take turns to play. Roll a dice three times to make a decimal number. The first roll is the units, the second is the tenths and the third is the hundredths.

Take your number away from 7·21. Let the other children check your subtraction.

Add the digits of your answer. Keep adding until you get a single-digit answer. Take that number of cubes. For example, if your answer was 3·56, you add 3 and 5 and 6 to get 14, add 1 and 4 to get 5 and take five cubes.

Play ten times. Who has the most cubes?

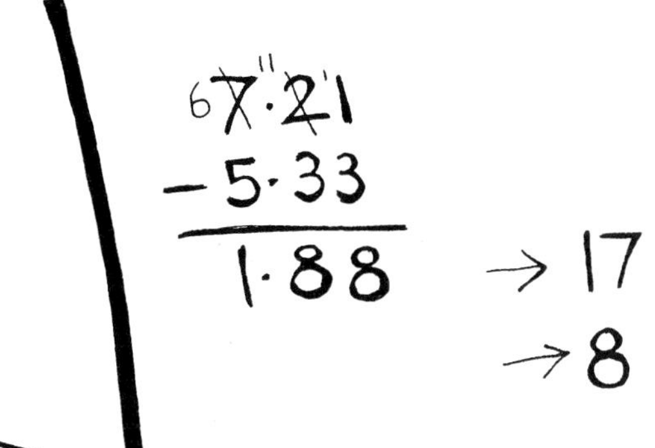

Abacus Ginn and Company 2001 Copying permitted for purchasing school only. This material is not copyright free.

N27 Properties of number

ACTIVITY 1
Whole class, in groups of 3

- *Listing factors of a given number*

Number cards (18, 24, 30, 36, 40, 44, 48, 56, 60) (PCMs 10 to 13), interlocking cubes, strips of paper
Shuffle the cards and place them in a pile face down. One child takes a card and writes the number at the top of a strip of paper. They pass the paper to the next child in their group who writes a factor of that number. That child passes it to the next child who writes another factor. This continues until all the factors have been written down. The child who writes the last factor takes a cube. Repeat with all the other cards. Who has the most cubes?

ACTIVITY 2
3–4 children

- *Creating numbers with a given number of factors*

Two dice (1 to 6), counters
Explain that one dice is for tens and the other is for units. The children take turns to roll both dice to make a 2-digit number. If they can make a number with exactly four factors, they collect a counter. The children check each other's numbers. Who is the first to collect 11 counters?

ACTIVITY 3
4 children

- *Finding factors of numbers*

Place-value cards (T, U) (PCMs 1 to 4), number cards (2 to 9) (PCM 9), counters
Each set of cards is shuffled and placed in two piles, face down. The children choose one card from each set to make a 2-digit number. They shuffle the number cards and deal out one each. If their card number is a factor of the 2-digit number they collect a counter. The children check each other's numbers. They deal out another card each and repeat. They then replace the number cards in the pile, take two new place-value cards and create a new 2-digit number. Repeat until all the place-value cards have been used. Who has the most counters?

ACTIVITY 4
3–4 children

- *Recognising pairs of factors of a given number*

Number cards (2 to 30) (PCMs 9 to 11), post-it notes
The cards are shuffled and placed face down in a pile. One child takes a card, shows it to the others and writes a pair of factors of that number on a post-it note. The second child writes another pair if they can, as does the third child. Continue until all the factors have been written on the note. Stick the note to the card. Repeat until all the cards have been used.

ACTIVITY 5
3–4 children

- *Finding factors of numbers*

Game 7: 'It's a Factor', counters, sets of coloured cubes, a dice
(See instructions on the card.)

GROUP ACTIVITY 6
3–4 children

- *Listing factors of a given number*

A dice (1 to 6), counters

GROUP ACTIVITY 7
2 pairs

- *Drawing rectangles for a given number, listing factors of a given number*

Number cards (0 to 9) (PCM 9), squared paper

Listing factors

Properties of number
GROUP ACTIVITY 6

N27

3–4 children

A dice (1 to 6), counters

Take turns to play. Roll the dice. If you roll 4, choose a number in column 4 and write all the factors of that number.

Let the others check. If you have listed all the factors, cover that number with a counter. If you have missed any, leave the number uncovered.

Continue playing until all the numbers are covered.

Dice numbers					
1	2	3	4	5	6
28	36	27	18	56	24
17	11	19	25	38	42
26	44	30	50	32	12
14	60	66	72	80	100

Rectangular factors

Properties of number
GROUP ACTIVITY 7

N27

2 pairs

Number cards (0 to 9), squared paper

In pairs, choose two number cards and create a 2-digit number.

Draw all the matching rectangles for that number on squared paper. The area of each rectangle must match the number.

Swap papers and numbers with the other pair.

List all the factors of that number.

Check that they have drawn a rectangle for every pair of factors.

Repeat with different 2-digit numbers.

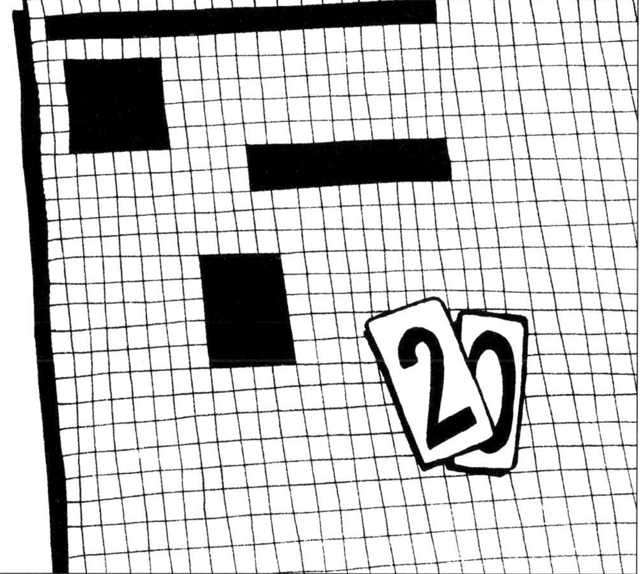

Abacus Ginn and Company 2001 Copying permitted for purchasing school only. This material is not copyright free.

N28 Properties of number

ACTIVITY 1
Whole class, in groups of 3–4

- *Recognising prime numbers and their adjacent numbers*

Number cards (1 to 60) (PCMs 9 to 13)

Shuffle the cards and place them in a pile face down. Each child takes five cards. One child puts down a prime number, e.g. 23. They take another card from the pile so that they still hold five cards. The next child may put down another prime number or may put an adjacent number next to the one already placed, e.g. they could put 22 or 24 next to 23. However, each prime number may only have one card on either side of it. If a child cannot put a card down, they still take a card from the pile. When all possible cards have been put down, who is holding the fewest cards?

ACTIVITY 2
3–4 children

- *Creating prime numbers*

Two dice (1 to 6), counters

The children decide which dice is for tens and which is for units. They take turns to roll both dice to make a 2-digit number. If they can make a prime number, they collect a counter. They check each other's numbers. Who is the first to collect 11 counters?

ACTIVITY 3
4 children

- *Creating prime numbers*

Place-value cards (T, U) (PCMs 1 to 4), counters

Each set of cards is shuffled and placed in two piles face down. Each child chooses one card from each set and makes a 2-digit number. If the number is a prime number they collect a counter. The children check each other's numbers. They replace the cards at the bottom of the piles, each take another card from each pile and repeat. Continue until one child has seven counters.

ACTIVITY 4
3 children

- *Creating a table to show prime numbers*

Multiplication square (1 to 100) (PCM 17), 5 by 6 grids

Each child writes the numbers 1 to 30 on their grid. They then consider each number in turn. If the number occurs only in the top row on the multiplication square, or not at all on the multiplication square, they colour it on their own grid. If the number is on the multiplication square, then they do not colour it. When they reach 30, they look at all the numbers they have coloured. Tell them that these are the prime numbers to 30.

GROUP ACTIVITY 5
3 children

- *Finding reduced numbers of prime numbers*

GROUP ACTIVITY 6
3 children

- *Recognising prime numbers*

Number cards (1 to 20) (PCMs 9, 10), post-it notes, counters

Prime reductions

Properties of number
GROUP ACTIVITY 5

3 children

Work together to list all the prime numbers up to 100.

Add the digits of each number, and keep adding, until you reach a 1-digit number for each one.

Do you notice a pattern?

Prime or not

Properties of number
GROUP ACTIVITY 6

3 children

Number cards (1 to 20), post-it notes, counters

Shuffle the cards and place them in a pile face down.

Take turns to play. Take a card. Multiply it by 6 and add or subtract 1.

Write your answer on a post-it note and stick the note on the card.

Look at the number on the note. Is this a prime number? If it is, take a counter.

Continue playing like this until all the cards have been used.

Who has the most counters?

Which cards have notes stuck to them which are **not** prime numbers?

19 × 6 = 114
114 + 1 = 115

Abacus Ginn and Company 2001 Copying permitted for purchasing school only. This material is not copyright free.

N29 Place-value

ACTIVITY 1
Whole class, in pairs

• *Creating a difference table involving positive and negative numbers*

Number cards (⁻10 to 10) (PCMs 8, 9)

On the board, draw a 5 × 5 grid for drawing up a difference table. Ask the children to copy it. Shuffle the cards and place them face down in a pile. Ask some children to select cards for the headings for the table: four column headings and four row headings. Write them in the grid and let the children copy the same arrangement. The children then complete the difference table, giving 16 differences. Check the results together.

ACTIVITY 2
2 children

• *Finding pairs of positive and negative numbers with a given difference*

Number cards (⁻10 to 10) (PCMs 8, 9), a dice (1 to 6)

Spread out the cards, face up. The children take turns to roll the dice and take a pair of cards which have a matching difference. They should check each other's pairs. If they cannot take a pair of cards, they miss a turn. The winner is the first to collect five pairs.

ACTIVITY 3
4 children

• *Recognising the difference between two positive and negative numbers, ordering positive and negative numbers*

Number cards (⁻10 to 10) (PCMs 8, 9)

The children shuffle the cards and deal out five each. Each child places their five cards in order, from smallest to largest. They look at the difference between adjacent pairs of numbers. Each child finds the pair which has the largest difference. This is their score. They check each other's scores and keep a score-sheet. The children play six rounds altogether and the winner is the child with the largest total score.

ACTIVITY 4
4 children

• *Recognising the difference between two positive numbers*

Playing cards (excluding the picture cards), interlocking cubes

The children shuffle the cards and deal out two each. Each child finds the difference between the two cards. The child with the largest difference collects a cube, as does the child with the smallest difference. They replace the cards, reshuffle and continue playing until one of them has collected eight cubes.

GROUP ACTIVITY 5
Pairs

• *Recognising the difference between two positive and negative numbers*

Number cards (⁻10 to 10) (PCMs 8, 9), counters (in two colours)

GROUP ACTIVITY 6
2–3 children

• *Ordering positive and negative numbers*

Number cards (⁻10 to 10) (PCMs 8, 9), a large sheet of paper, interlocking cubes

N29

Find the difference

Place-value
GROUP ACTIVITY 5

Pairs

Number cards (⁻10 to 10), counters (in two colours)

Shuffle the cards and place them face down in two equal piles.

In turn, take a card from each pile, then find the difference between the numbers. If the answer is below, cover it with one of your counters. Check each other's calculation. Use the number line to help.

Continue until the winner has four counters covering numbers.

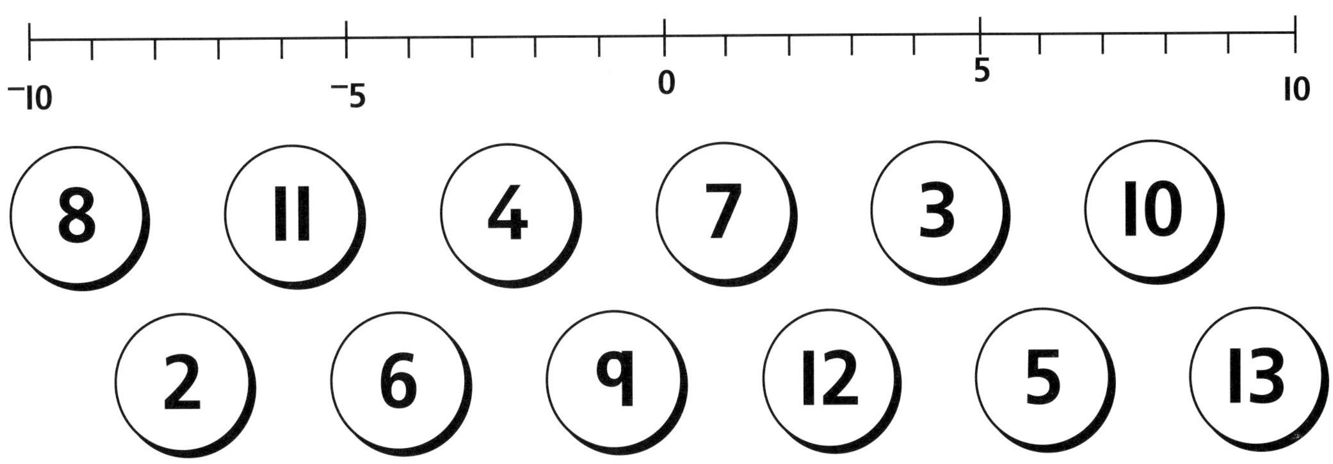

N29

Five in order

Place-value
GROUP ACTIVITY 6

2–3 children

Number cards (⁻10 to 10), a large sheet of paper, interlocking cubes

Draw a line of five boxes, each large enough to contain a number card:

Shuffle the cards and place them in a pile, face down.

In turn, reveal five cards, one at a time, placing them in the boxes. Your aim is to get all five in order, from smallest to largest, left to right.

After each card is revealed, you must place it in a box, and not move it. If your final order is correct, collect a cube.

Play several rounds each to see who collects the most cubes.

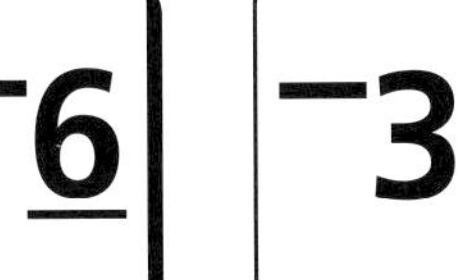

Abacus Ginn and Company 2001 Copying permitted for purchasing school only. This material is not copyright free.

N30 Place-value

ACTIVITY 1
Whole class, in pairs

- *Estimating a multiplication, U × HTU, by rounding*

Place-value cards (H, T, U) (PCMs 1 to 4), a dice (1 to 6)

Shuffle the cards separately and place them face down in three piles. One child takes a card from each pile to create a 3-digit number, e.g. 387. Another child rolls the dice, e.g. 4. Write a multiplication of the two numbers on the board, i.e. 4 × 387. The children round the 3-digit number, and use this to approximate the answer. They then multiply the numbers accurately, starting with the hundreds, then the tens, and finally the units, adding the three answers together. They find the difference between the estimation and the accurate answer. Repeat several times.

ACTIVITY 2
3–4 children

- *Estimating a multiplication, TU × TU, by rounding*

Number cards (30 to 80) (PCMs 11 to 14), a calculator, interlocking cubes

The children shuffle the cards and turn over two. All the children estimate the answer. They use the calculator to check the answer accurately. The child whose estimate is the closest collects a cube. If two players tie, they both collect a cube. Repeat for ten rounds. Who collects the most cubes?

ACTIVITY 3
4 children

- *Estimating a multiplication, U·th × U, by rounding*

Place-value cards (U, t, h) (PCMs 3 to 6), number cards (2 to 9) (PCM 9), a calculator, interlocking cubes

The children shuffle the place-value cards and turn over one of each to create a decimal number, e.g. 7·48. They shuffle the number cards and deal out a multiplier, e.g. 4. All the children estimate the product. They use the calculator to check the answer accurately. The child whose estimate is the closest collects a cube. If two players tie, they both collect a cube. Repeat for ten rounds. Who collects the most cubes?

ACTIVITY 4
3–4 children

- *Estimating the perimeter and area of rectangles by approximating*

Different-sized card rectangles (dimensions not in exact centimetres), a ruler

Each child writes an estimate of the perimeter and area of each rectangle by looking at them. Next, using a ruler they accurately measure the dimensions of each rectangle. They make new estimates of the perimeter and area based on these measurements. Finally, they calculate the perimeter with paper and pencil, and calculate the area using a calculator. They check their results to see how close their estimates were.

GROUP ACTIVITY 5
Pairs

- *Estimating a multiplication, U·t × U, by rounding*

A calculator, scissors

GROUP ACTIVITY 6
Pairs

- *Estimating a division, HTU ÷ TU, by rounding*

A calculator

N30

All in order

Pairs

Place-value
GROUP ACTIVITY 5

A calculator, scissors

Cut out the 12 rectangles and arrange them in an estimated order. Place the number which you think has the smallest answer on the left, then the next smallest, and so on, with the estimated largest answer on the right.

Estimate each answer by rounding the decimal number.

Next, calculate each answer, writing it in the square.

Check your estimated order.

How many multiplications were out of place?

4·9 × 3	7·8 × 4	6·2 × 7
5·3 × 9	4·6 × 8	14·4 × 3
17·4 × 2	5 × 7·4	8 × 6·9
7 × 7·5	6 × 8·3	4 × 10·6

N30

Division estimation

Pairs

Place-value
GROUP ACTIVITY 6

A calculator

Estimate each answer by rounding the 2-digit numbers.

Write the estimate in the table.

Use a calculator to work out each division, rounding the answer on the calculator to the nearest whole number.

Find the difference between each of these and the estimates. How close are your estimates?

Division	Estimate	Accurate	Difference
416 ÷ 19			
386 ÷ 21			
862 ÷ 39			
742 ÷ 19			
573 ÷ 11			
486 ÷ 31			

Abacus Ginn and Company 2001 Copying permitted for purchasing school only. This material is not copyright free.

N31 Multiplication/division

ACTIVITY 1
Whole class, in pairs

- *Multiplying a 1-place decimal number by a 1-digit number*

Place-value cards (U, t) (PCMs 3 to 5), a dice (1 to 6), counters

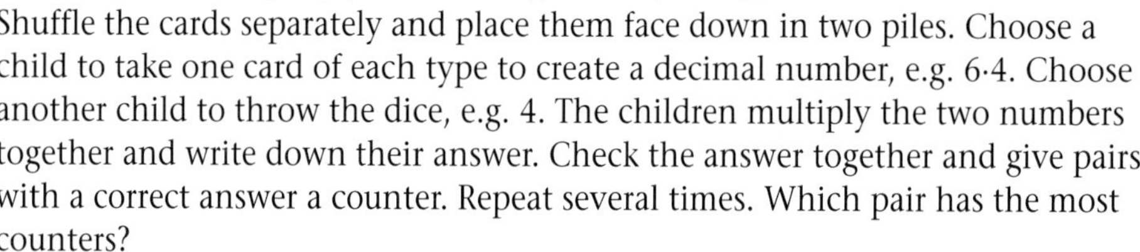

Shuffle the cards separately and place them face down in two piles. Choose a child to take one card of each type to create a decimal number, e.g. 6·4. Choose another child to throw the dice, e.g. 4. The children multiply the two numbers together and write down their answer. Check the answer together and give pairs with a correct answer a counter. Repeat several times. Which pair has the most counters?

ACTIVITY 2
3–4 children

- *Estimating the result of multiplying a 1-place decimal number by a 1-digit number*

Place-value cards (U, t) (PCMs 3 to 5), number cards (1 to 9) (PCM 9), a calculator, counters

The cards are shuffled separately and placed face down in three separate piles. The children take two place-value cards to make a decimal number, and a number card for the multiplier. Each child rounds the decimal number to its nearest whole number, then multiplies it to estimate the answer. They check each other's rounding and estimates. One child multiplies the numbers together using the calculator. How close are their estimates? Repeat for different multiplications, with the children sharing roles.

ACTIVITY 3
2 pairs

- *Multiplying a 1-place decimal number by a 1-digit number*

Place-value cards (U, t) (PCMs 3 to 5), number cards (2 to 9) (PCM 9), a calculator, cubes

The cards are shuffled separately and placed face down in three piles. Each pair takes two place-value cards, one of each type to create a decimal number, and also a number card. They secretly multiply the two numbers together. They then exchange cards and multiply the other two numbers together. They check their results to see if they match. They can use a calculator if there is a dispute. The pairs collect a cube for a correct answer. Repeat several times. Which pair collects the most cubes?

ACTIVITY 4
2–3 children

- *Multiplying a 1-place decimal number by a 1-digit number, creating multiplications which are close to a given product*

Number cards (1 to 9) (PCM 9), counters

The children select three cards to make a 1-place decimal number, using a counter for the decimal point, and a 1-digit number. They then multiply them together. They investigate which choice of cards and numbers will give a product which is closest to 10, 15, 20, 25, ... 80.

GROUP ACTIVITY 5
2–3 children

- *Multiplying a 1-place decimal number by a 1-digit number*

A calculator, counters, a one-minute timer, cubes

GROUP ACTIVITY 6
3 children

- *Multiplying a 1-place decimal number by a 1-digit number, creating multiplications which are close to a given product*

Number cards (1 to 9) (PCM 9), a calculator

N31

Decimal multiplication

Multiplication/division
GROUP ACTIVITY 5

2–3 children

A calculator, counters, a one-minute timer, interlocking cubes

Cover each star with a counter.

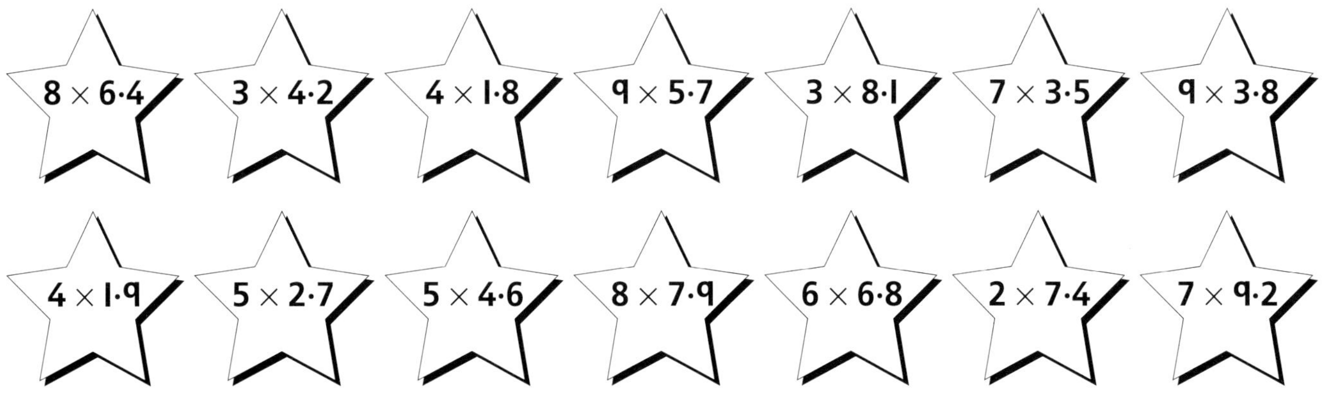

Take turns to remove a counter. You each have one minute, using the timer, to do the calculation.

Check the results together, using a calculator if necessary. If you are correct, collect a cube.

When all counters have been removed, the winner is the player with the most cubes.

N31

Target multiplication

Multiplication/division
GROUP ACTIVITY 6

3 children

Number cards (1 to 9), a calculator

Each make a scoresheet like the one below.

Shuffle the cards and deal three each.

With the cards, each create a multiplication of a 1-place decimal number by a 1-digit number, aiming for an answer as close as possible to one of the target numbers.

Multiply the numbers together, writing them on the scoresheet.

Check each other's answers, using a calculator if necessary.

Find the difference between the answer and the target.

The winner of the round is the player whose answer is the closest to the target.

Target	Multiplication	Difference
25 15 30 10 40		

Target | Multiplication | Difference
25 | 3 × 9·1 = 27·3 | 2·3

Abacus Ginn and Company 2001 Copying permitted for purchasing school only. This material is not copyright free.

N32 Multiplication/division

ACTIVITY 1
Whole class, in pairs

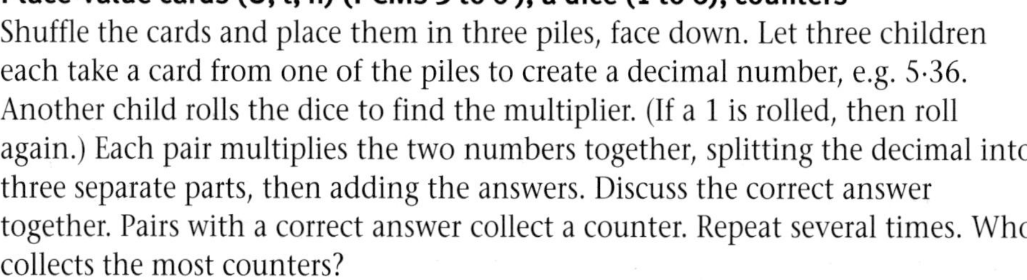

- *Multiplying U × U·th, using standard written methods*

Place-value cards (U, t, h) (PCMs 3 to 6), a dice (1 to 6), counters
Shuffle the cards and place them in three piles, face down. Let three children each take a card from one of the piles to create a decimal number, e.g. 5·36. Another child rolls the dice to find the multiplier. (If a 1 is rolled, then roll again.) Each pair multiplies the two numbers together, splitting the decimal into three separate parts, then adding the answers. Discuss the correct answer together. Pairs with a correct answer collect a counter. Repeat several times. Who collects the most counters?

ACTIVITY 2
3–4 children

- *Multiplying U × U·th, using standard written methods*

Place-value cards (U, t, h) (PCMs 3 to 6), a calculator, counters
Play 'Multiplying by 2'. The children shuffle the cards in separate piles and turn over three to make a decimal number. They multiply the number by 2. They check each other's answers, using a calculator if necessary, and collect a counter if they are correct. They put the cards back into the pack, turn over three more and play 'Multiplying by 3'. They continue up to multiplying by 9. Who has the most counters?

ACTIVITY 3
2–3 children

- *Multiplying U × U·t, using informal written methods*

Place-value cards (U, t) (PCMs 3 to 5), a dice (1 to 6), counters
The children shuffle both sets of cards separately and place them in two piles, face down. They take turns to reveal one card from each pile to make a decimal number, then throw the dice for a multiplier. (If a 1 is rolled, then roll again.) All children do the calculation by drawing a rectangular grid and splitting the decimal number into two parts: units and tenths. They write the answer to each part inside the grid, then combine the answers. They check each other's calculations. Those with a correct answer collect a counter. Repeat several times. Who has the most counters?

ACTIVITY 4
Pairs

- *Estimating the result of multiplying U × U·th, multiplying U × U·th, using informal or standard written methods*

Number cards (1 to 9) (PCM 9), a calculator, counters
The children deal out four cards each. They both arrange three of them to make a 2-place decimal number, using a counter between them for the decimal point. The other card is the multiplier. Each child multiples their two numbers together and they check each other's multiplication. The child whose answer is closest to 20 takes a counter. They replace the cards and repeat several times. Who is the first to collect five counters? Extend the activity by changing the target number.

GROUP ACTIVITY 5
3–4 children

- *Multiplying U·th × U, using standard written methods*

Place-value cards (U, t, h) (PCMs 3 to 6), counters

GROUP ACTIVITY 6
3–4 children

- *Multiplying U·th × U, using informal written methods*

Number cards (1 to 9) (PCM 9)

Multiplying decimals

Multiplication/division
GROUP ACTIVITY 5

3–4 children

Place-value cards (U, t, h), counters

Choose a 1-digit multiplying number, e.g. 3.

Deal five cards from each set and make five different 2-place decimal numbers.

In turn, each multiply the decimal numbers by 3.

Check each other's answers. Collect a counter for each correct answer.

Repeat five times, choosing different multiplying numbers and making different decimal numbers each time.

Who collects the most counters?

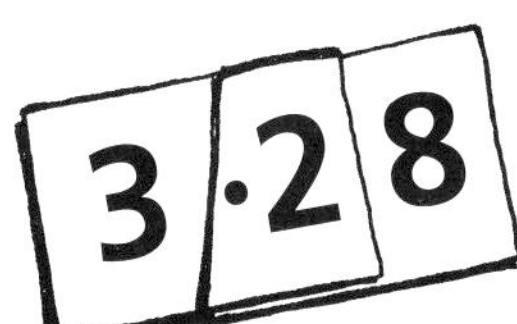

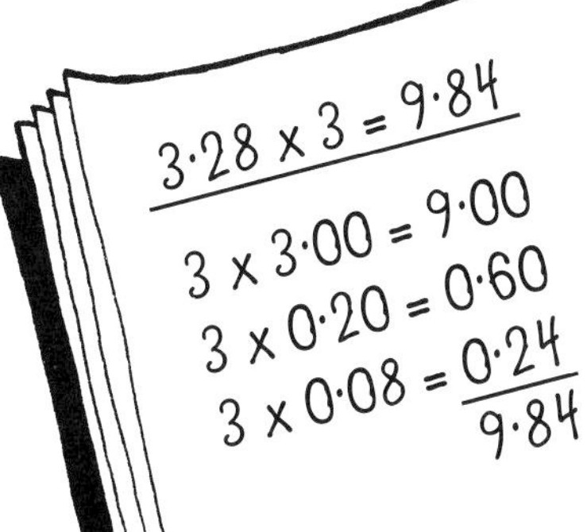

Card decimals

Multiplication/division
GROUP ACTIVITY 6

3–4 children

Number cards (1 to 9)

Deal out three cards to make a 2-place decimal number.

Deal another card for a multiplier.

Each write a multiplication of the decimal number by the card number.

Check each other's answers.

Shuffle the cards and repeat several times.

Abacus Ginn and Company 2001 Copying permitted for purchasing school only. This material is not copyright free.

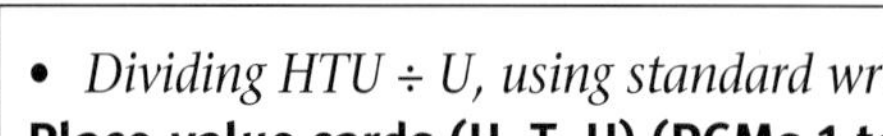

N33 Multiplication/division

ACTIVITY 1
Whole class, in pairs

- *Dividing HTU ÷ U, using standard written methods*

Place-value cards (H, T, U) (PCMs 1 to 4)

Shuffle the cards in three separate piles, face down. Three children take a card from one of the piles to create a 3-digit number. All the children divide the number by 2, then 3, then 4, ... up to 10, expressing each remainder as a fraction. When dividing by 10, they should express the remainder as a decimal too. The children check each other's answers, scoring one point for each correct one. Who collects the most points?

ACTIVITY 2
2–3 children

- *Dividing HTU by a multiple of 10, using standard written methods*

A holiday brochure

The children study the brochure. Each child finds five holidays that they think they will enjoy and writes down their price. The children work together to find out how much each holiday would cost per week if they saved for it over 20 weeks. For example, the holiday costs £439 which gives $439 \div 20 = 21 \text{ r } \frac{19}{20}$, or nearly £22 per week. Which one would they choose?

ACTIVITY 3
4 children

- *Dividing HTU ÷ U, using standard written methods*

A dice (1 to 6), interlocking cubes

One child rolls a dice three times to make a 3-digit number – the first roll for the hundreds, the second for the tens, and the third for the units. One child divides the number by 2, another by 3, another by 4, and another by 5. If the number divides exactly, they collect a cube. The children check each other's work. Repeat for ten more numbers, with each child dividing by a different number up to 9 every time. Who collects the most cubes?

ACTIVITY 4
3–4 children

- *Dividing HTU ÷ TU, using standard written methods, estimating the result of dividing HTU ÷ TU*

Place-value cards (H, T, U) (PCMs 1 to 4), number cards (11 to 20) (PCM 10), counters

The children separately shuffle the three sets of place-value cards, and turn over one from each set to make a 3-digit number. They shuffle the number cards and take one for the divider. They estimate the result of dividing one by the other, and show their written estimate. Each child then works out the division accurately and they check each other's work. They discuss who has the closest estimate and that child collects a counter. They put the cards back into the pack and repeat several times. Who has the most counters?

GROUP ACTIVITY 5
3 children

- *Dividing HTU ÷ TU, using standard written methods*

Place-value cards (H, T, U) (PCMs 1 to 4), counters (three colours)

GROUP ACTIVITY 6
2–3 children

- *Dividing HTU by a multiple of 10, using standard written methods*

Number cards (2 to 9) (PCM 9), cubes

Spot numbers

Multiplication/division
GROUP ACTIVITY 5

N33

3 children

Place-value cards (H, T, U), counters (three colours)

Deal out one of each type of place-value card to create a 3-digit number, e.g. 468.

Each choose two of the six spot numbers below by placing your own counters in the spots.

Each divide the 3-digit number by each of the spot numbers, in turn.

Record the size of the remainder for each division.

If your dividing number gives the largest remainder, you win.

Play again, with a different 3-digit number.

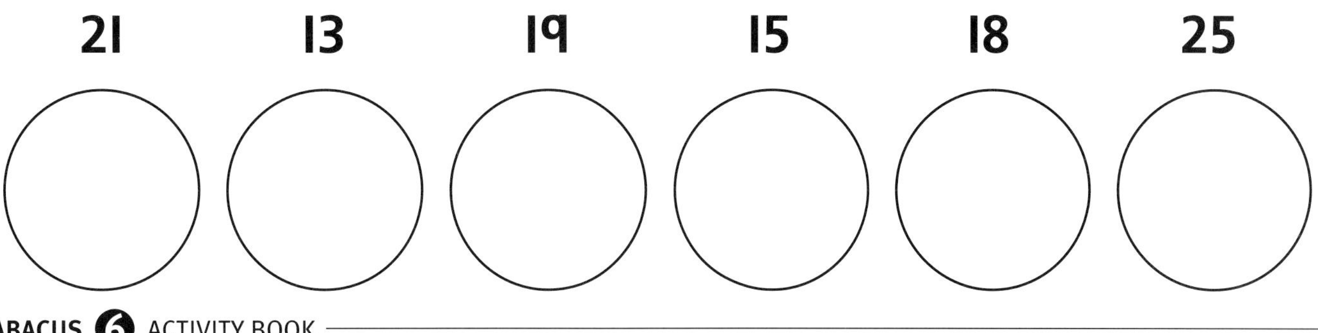

Divide into 720

Multiplication/division
GROUP ACTIVITY 6

N33

2–3 children

Number cards (2 to 9), cubes

Shuffle the cards and place them face down in a pile.

Take turns to take a card, multiply it by 10 and write the answer.

Each divide 720 by your number. Check each other's work.

Collect cubes to match the answer.

If you cannot divide exactly, take two cubes.

Replace the cards at the bottom of the pile and play again.

Continue until one player has 20 cubes.

$720 \div 60 = 12$

Abacus Ginn and Company 2001 Copying permitted for purchasing school only. This material is not copyright free.

N34 Multiplication/division

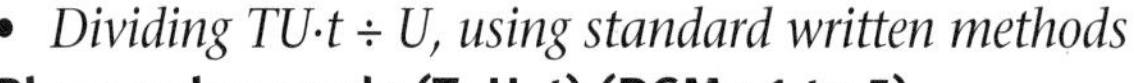

ACTIVITY 1
Whole class, in pairs

- *Dividing TU·t ÷ U, using standard written methods*

Place-value cards (T, U, t) (PCMs 1 to 5)
Shuffle the cards into three piles, face down. Three children take a card from one of the piles to create a 1-place decimal number. The children divide the number by 2, then 3, then 4, ... up to 10, giving the answers to one decimal place. When dividing by 10, they should express the remainder as a decimal. The children check each other's answers, scoring one point for each correct one. Who collects the most points?

ACTIVITY 2
2–3 children

- *Dividing U·th ÷ U, using standard written methods*

A book or toy catalogue, post-it notes
The children study the catalogue and choose a book or toy they particularly like. They work out how much they would pay if they each paid one third of the price, e.g. the book costs £5·49, so £5·49 ÷ 3 = £1·83. They write the answer on a post-it note and stick it in the catalogue next to the item. Repeat at least ten times.

ACTIVITY 3
4 children

- *Dividing U·t ÷ U, using standard written methods*

A dice (1 to 6), interlocking cubes
One child rolls a dice twice to make a 1-place decimal number – the first roll for the units, the second for the tenths. One child divides the number by 2, another by 3, another by 4, and another by 5. If the number divides exactly, the children collect a cube. They check each other's work. Repeat for ten more numbers, with each child dividing by a different number each time. Who collects the most cubes?

ACTIVITY 4
3–4 children

- *Dividing TU·t ÷ U, using standard written methods, estimating the result of dividing TU·t ÷ U*

Place-value cards (T, U, t) (PCMs 1 to 5), number cards (2 to 9) (PCM 9), counters
The children separately shuffle the three sets of place-value cards, and turn over one from each set to make a 1-place decimal number. They shuffle the number cards and take one for the divider. The children estimate the result of dividing one by the other, and show their written estimate. Each child then completes the division accurately. They check each other's work and discuss whose estimate is the closest. That child collects a counter. They replace the cards and repeat several times. Who has the most counters?

GROUP ACTIVITY 5
3 children

- *Dividing TU·t ÷ U, using standard written methods*

Place-value cards (T, U, t) (PCMs 1 to 5), interlocking cubes, a dice (1 to 6)

GROUP ACTIVITY 6
3 children

- *Dividing TU·t ÷ U, using standard written methods*

Interlocking cubes

Dividing decimals

Multiplication/division
GROUP ACTIVITY 5

3 children

Place-value cards (T, U, t), interlocking cubes, a dice (1 to 6)

Shuffle the cards separately and place them face down in three piles.

Take turns to take a card from each pile and make a decimal number.

Roll the dice and divide your card number by the dice number.

Write down the division.

After one round check each other's work.

If your answer is correct, round it to the nearest whole number and take a matching number of cubes.

Play again, until all the cards are taken.

Who has the most cubes?

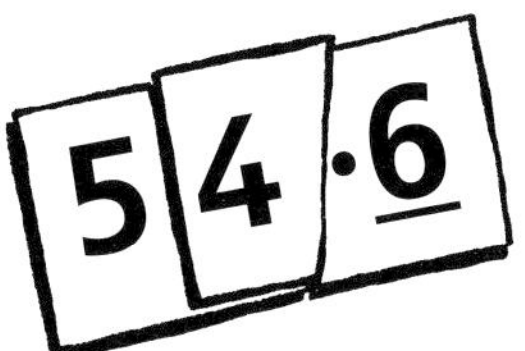

What target?

Multiplication/division
GROUP ACTIVITY 6

3 children

Interlocking cubes

Each choose a target number. It should be between 10 and 50 and have tenths, for example, 35·6.

Divide it by any number from 2 to 9. Write down the answer.

Check each other's work.

Look at your remainder. Take a matching number of cubes.

Play ten rounds, choosing a different target number each time.

Who collects the most cubes?

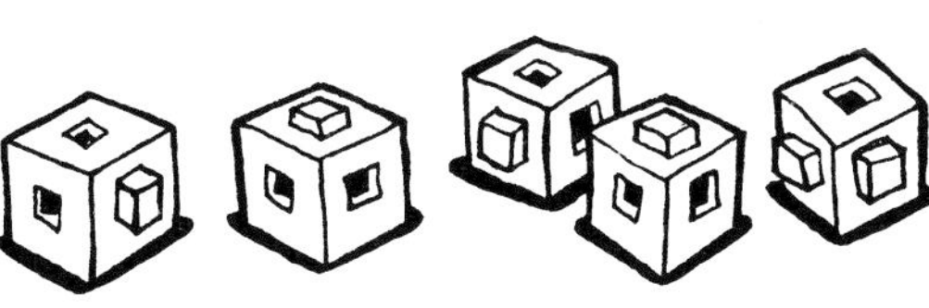

Abacus Ginn and Company 2001 Copying permitted for purchasing school only. This material is not copyright free.

N35 Percentages

ACTIVITY 1
Whole class

- *Calculating percentage reductions*

A toy or clothes catalogue

Ask a child to look at the catalogue and choose an item that they would like to buy. Write the price on the board. The children work out how much the item would cost if the price was reduced by 15%. They do this by working out a 10% reduction, halving that to find 5% and adding the two together. They then work out the new price. Check that they all have the correct answer. Repeat, asking another child to choose an item.

ACTIVITY 2
Pairs

- *Calculating percentage increases and reductions*

A current TV guide, post-it notes

The children study the TV guide together. They choose six programmes that they enjoy watching and write the start and finish time of each one. They calculate how long each programme would be if it was 25% longer. They write the old duration and the extended one on a post-it note and stick it next to the programme. The children then choose six programmes that they do not like. They work out how long these would be if they were all 10% shorter.

ACTIVITY 3
2 pairs

- *Calculating the percentage of a given amount of money*

Coins (£1, 10p), post-it notes, a calculator

One pair puts out an amount of money using the £1 and 10p coins. The second pair works out 10% of the amount. They write the amount on a post-it note and stick it beside the money. The first pair checks the calculation to see if they agree. The children make a further check, using the calculator. If necessary, they write the correct amount below the first amount. Repeat at least ten times.

ACTIVITY 4
3 children

- *Calculating the percentage to one decimal place of a given amount of money*

A local newspaper or car catalogue

The children study the newspaper or catalogue together. They choose a car they like and work out the VAT at 17·5% on that car price. Repeat six times.

GROUP ACTIVITY 5
Pairs

- *Calculating percentage reductions*

A travel brochure, strips of paper

GROUP ACTIVITY 6
Pairs

- *Calculating percentage reductions and increases, calculating the difference between prices*

Ticket to ride

Percentages
GROUP ACTIVITY 5

Pairs

Travel catalogue, strips of paper

Make an aeroplane ticket to a place you would both like to visit. Write a realistic price on the ticket.

Each calculate the price of the ticket if it was on special offer with a 15% reduction.

What would it cost if the reduction was 25%?

Write down the new prices on the tickets.

Repeat six times, choosing a different destination each time.

£150
10% = £15
5% = £ 7.50
15% = £22·50
£150 – £22·50 =

CD price war

Percentages
GROUP ACTIVITY 6

Pairs

Write the titles of two CDs you both like to listen to.

Write a price for each one, making sure they are different.

Find the cost if:

- the CD is reduced in price by 25%
- the CD is raised in price by 25%.

Find the difference between these two new prices.

Repeat, this time reducing and raising the price by 15%.

£12·80
25% = £3·20
£12·80 – £3·20
= £9·60

Abacus Ginn and Company 2001 Copying permitted for purchasing school only. This material is not copyright free.

N36 Fractions/decimals

ACTIVITY 1
Whole class

- *Writing percentages as equivalent fractions*

Percentage cards (10%, 20%, 25%, 30%, 33%, 40%, 50%, 60%, 66%, 70%, 75%, 80%, 90%) (PCM 20), post-it notes

The cards are shuffled and spread out face down. Ask a child to turn over a card and look at it secretly. They write an equivalent fraction on a post-it note and stick it on the back of the card. Another child reads the fraction and says the percentage on the card. If they are correct, they turn over the next card. Play until all the cards have been used.

ACTIVITY 2
3 children

- *Converting fractions to decimals*

A calculator, post-it notes

One child writes a fraction in its simplest form on a post-it note, e.g. $\frac{2}{5}$. The other two children work out what this fraction is when it is expressed as a decimal. They may use a calculator to help them if necessary. They write the equivalent decimal number on the note. Repeat at least ten times, with the children taking different roles.

ACTIVITY 3
3 children

- *Converting percentages to actual amounts*

Interlocking cubes, post-it notes

Using the cubes, each child builds a model. Tell them that 20% of the model must be made with red cubes, 10% with blue cubes and 25% with green. The rest can be with any other colours. When they have built their models, they write the total number of cubes and the numbers of red, blue and green cubes. Are their percentages correct?

ACTIVITY 4
3 children

- *Converting fractions to decimals*

A calculator, post-it notes

Using the post-it notes, the children write a series of related fractions in their simplest forms, e.g. $\frac{1}{6}, \frac{1}{3}, \frac{2}{3}, \frac{5}{6}$ or $\frac{1}{8}, \frac{1}{4}, \frac{3}{8}, \frac{5}{8}, \frac{3}{4}, \frac{7}{8}$. They then convert each fraction into its decimal equivalent, using a calculator if necessary. What patterns do they notice?

GROUP ACTIVITY 5
3 children

- *Calculating percentages and fractions of given numbers, subtracting a smaller number from a larger one*

Counters

GROUP ACTIVITY 6
Pairs

- *Calculating percentages*

A picture book with words

Cover up

Fractions/decimals
GROUP ACTIVITY 5

3 children

Counters

Take turns to play.

Choose two amounts on the grid below. Work out the difference between the two amounts. This is your score. If the two amounts are identical, cover them with counters.

Continue until all the spaces on the grid except one are covered.

The player with the lowest score is the winner.

20% of 500	$\frac{3}{4}$ of 20	$\frac{1}{2}$ of 60	$\frac{2}{3}$ of 120	25% of 36
10% of 4	30% of 50	50% of 15	75% of 20	$\frac{1}{3}$ of 1·2
$\frac{1}{4}$ of 40	6% of 50	$\frac{2}{5}$ of 250	3% of 300	$\frac{1}{3}$ of 45
$\frac{1}{10}$ of 10	10% of 90	30% of 30	1% of 1000	10% of 75
15% of 200	2% of 2500	10% of 800	1% of 100	25% of 200

Letter percentage

Fractions/decimals
GROUP ACTIVITY 6

Pairs

A picture book with words

Look at the first page of the book. Find a letter that is quite common, for example, 'e'.

Count every one of these letters.

Count or estimate accurately the total number of letters on the page.

Work out the percentage of your chosen letter on that page.

Do the same on other pages. Is the percentage about the same?

Choose another letter and repeat the activity.

Abacus Ginn and Company 2001 Copying permitted for purchasing school only. This material is not copyright free.

N37 Ratio/proportion

ACTIVITY 1
Whole class, in pairs

• *Calculating proportion*

Strips of squared paper (24 squares long)

Ask each child to colour a pattern on their strip of paper, using three colours. When they have finished, they exchange strips. They work out the proportion of each colour and write it on the back of the strip. For example, 4 out of 24 squares are red, so $\frac{4}{24}$ or $\frac{1}{6}$ are red. Each child then checks their own strip to check that the proportions are correct and that they are expressed in the simplest form. Repeat with a new strip of paper.

ACTIVITY 2
Pairs

• *Calculating proportion*

A current TV guide

The children study the TV guide. Choosing one day between 6 p.m. and midnight, they calculate the number of minutes which are devoted to news or news-type programmes. They then calculate the proportion of time given to news programmes in the whole evening's viewing. They then do the same for another evening. Is the proportion the same? Extend the activity, by asking them to find the proportion of the evening's viewing which is devoted to soaps.

ACTIVITY 3
Pairs

• *Calculating proportion*

Interlocking cubes

The children build a long strip of 60 cubes in a pattern of three colours. They count the number of each colour and write each as a proportion of the whole. For example, there are 24 red cubes so the proportion of red is $\frac{24}{60}$. The children then discuss whether this proportion can be written in a simpler form, i.e. $\frac{24}{60}$ is $\frac{2}{5}$.

GROUP ACTIVITY 4
Pairs

• *Calculating proportion*

A picture book with words

GROUP ACTIVITY 5
Pairs

• *Calculating proportion*

A class register

Word proportions

Ratio/proportion
GROUP ACTIVITY 4

Pairs

A picture book with words

Look at the first page of the book. Count the words on the page.

Count how many words there are which have less than four letters.

Work out the proportion of those short words on that page.

Do the same with other pages. Is the proportion about the same?

Repeat this activity, looking for a particular word, such as 'the'.

Number of words = 85

Name proportions

Ratio/proportion
GROUP ACTIVITY 5

Pairs

A class register

Write a list of the first names of the children in the class.

Count how many there are.

Count how many names begin with a vowel.

Work out the proportion of names in the class that begin with a vowel.

Do the same for names beginning with a certain letter, such as 'S'.

Abacus Ginn and Company 2001 Copying permitted for purchasing school only. This material is not copyright free.

N38 Ratio/proportion

ACTIVITY 1
Whole class, in pairs

- *Calculating ratio*

Strips of squared paper (24 squares long)

Ask the children to colour some squares on their strip of paper. Explain that the ratio of coloured to non-coloured squares must be 3:1, that is, three coloured squares to one non-coloured square. The children then swap strips and check that the ratio is correct. On the reverse of the strip, they write the number of squares coloured and not coloured. The children then do the same, with strips 20 squares long. Before they start to colour, the children should write down the number of squares that they will not be colouring.

ACTIVITY 2
2–3 children

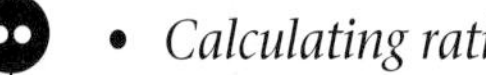

- *Calculating ratio*

Several music CDs

The children each select a CD and count the number of tracks on it. They then count the number of tracks which are less than three minutes long and those that are over three minutes long. They write the ratio of tracks over three minutes to tracks under three minutes. The children compare notes. They then do the same with three more CDs. Is the ratio constant or does it change? To obtain a clearer picture, they should do the same with three more CDs. They then write an average ratio of tracks over three minutes to tracks under three minutes. Finally, they work out how many tracks are likely to last less than three minutes on a 30 track CD.

ACTIVITY 3
3 children

- *Illustrating ratio*

Three grids (4×5, 5×6, 3×5)

The children take a grid each and colour it in a ratio 3:2, red to yellow. They then count the number of red squares and write it on the back of the grid and do the same for the number of yellow squares. Finally, they write these two numbers as a ratio, with the ratio 3:2 below it. Extend the activity by using three more grids (3×4, 2×6, 6×4) and colouring in a ratio of 2:4.

GROUP ACTIVITY 4
Pairs

- *Calculating amounts of money based on a ratio of coins*

Coins (10p, 20p, 50p, £1), a cloth bag

GROUP ACTIVITY 5
Pairs

- *Calculating ratio*

A class register

Money ratios

Pairs

Ratio/proportion
GROUP ACTIVITY 4

N38

Coins (10p, 20p, 50p, £1), a cloth bag

Agree a ratio, e.g. 2:3 and write it down.

Decide which two types of coin you will both take from the bag, and take them in the ratio agreed. Tell each other how much money you have.

Each work out what coins your partner has taken.

Check to see if you are both correct. Write down the ratios, and the amounts.

Do the activity again, starting with a new ratio and choosing two different coins.

Gender ratios

Pairs

Ratio/proportion
GROUP ACTIVITY 5

N38

A class register

Count the numbers of boys and girls in your class.

Write the numbers as a ratio, for example, 14:16, boys to girls. Is there a simpler way of writing this ratio? For example, 7:8.

If this ratio is already in its simplest form, calculate the numbers of boys and girls in classes of between 20 and 60 children with the same ratio of boys to girls.

Now work out the ratio of 10 year-olds to 11 year-olds in your class.

Abacus Ginn and Company 2001 Copying permitted for purchasing school only. This material is not copyright free.

Addition/subtraction

N39

ACTIVITY 1
Whole class, in pairs

- *Mentally adding 2-digit numbers and checking the total in different ways*

Number cards (20 to 80) (PCMs 10 to 14)

One child chooses four cards. The other adds the numbers in their head, writing them in the order they want to add them. For example, Sam chooses 34, 60, 29 and 56. Tom writes them like this: 60 + 56 + 29 + 34 = 179. The first child checks the addition by adding the numbers in a different order or by using a different method (e.g. vertical addition). They must demonstrate that they have used a different calculation. For example, Tom checks Sam's work by adding 60 + 34 = 94 and 56 + 29 = 85, so 94 + 85 = 179. Repeat ten times, with the children swapping roles.

ACTIVITY 2
3–4 children

- *Checking subtractions by adding on*

Number cards (2 to 100) (PCMs 9 to 15), post-it notes, interlocking cubes

The children shuffle the cards and deal out five cards each. Each child puts down two cards and says the difference between them. They write their difference on a post-it note and stick it beside the cards. They take two more cards each and do the same. They continue until one of them has a total of differences which is a multiple of 10. At this point, they stop taking cards. Each child checks the subtractions of their neighbour. They do this by adding one of the card numbers to the difference written on the post-it note in each pair of cards. For example, if the cards are 34 and 61 and the note says 27, they add 27 to 34 to check that this comes to 61. Any wrong answers must be corrected and the child who made the mistake must take a cube. Continue until everyone has done at least 20 subtractions. The winner is the child with the least cubes.

ACTIVITY 3
Pairs

- *Adding and subtracting 2-digit numbers in the quickest way*

Number cards (20 to 80) (PCMs 10 to 14), a calculator, post-it notes

The children shuffle the cards and place them in a pile face down. One child turns over three cards. Each child has to add the two largest numbers and subtract the smallest number as quickly as they can. One of them must do it in their head, while the other has to use a calculator. They each write their answer on a post-it note. They compare answers. If they are not the same, they check the calculation, using a different strategy. The child who found the correct answer quickest scores ten points. Repeat at least ten times with the children taking different roles. Who has the most points?

GROUP ACTIVITY 4
3 children

- *Adding and subtracting amounts of money*

Coins (1p, 2p, 5p, 10p, 20p, 50p, £1), a cloth bag, interlocking cubes

GROUP ACTIVITY 5
Pairs

- *Adding amounts of money*

Coins (20p, 50p)

Money bags

Addition/subtraction
GROUP ACTIVITY 4

3 children

Coins (1p, 2p, 5p, 10p, 20p, 50, £1), a cloth bag, interlocking cubes

One of you take an amount of money from the bag. Choose a purse by pointing to it.

Compare the amount you took with the amount in that purse and write down the difference and the total.

The other two children decide how much money was taken. They should check your arithmetic. Are your calculations correct? If not, you take a cube.

Put the money back in the bag and play ten more times, sharing the roles.

Who has the fewest cubes?

Grid conundrum

Addition/subtraction
GROUP ACTIVITY 5

Pairs

Coins (20p, 50p)

Draw a large grid as shown. One of you choose a 20p coin, the other a 50p coin. Place your coin on one of the 'Start' squares.

The aim is to get across the grid. You may move horizontally or vertically, one square per turn. Each time you land on a square, score the number on the square multiplied by your coin.

Write down your score each turn.
After two moves, check each other's arithmetic.

Use a different method to check again.
Keep playing, until you have both reached the opposite corner.

Calculate your total scores.

You win if your score is nearer to a multiple of £50 than your partner's.

Swap coins and play again.

Start	22	30	15
26	36	45	80
44	38	76	12
32	50	19	Start

Abacus Ginn and Company 2001 Copying permitted for purchasing school only. This material is not copyright free.

N40 Addition/subtraction

ACTIVITY 1
Whole class, in pairs

• *Repeatedly adding 25 to a given number*

Strips of paper

Give the children a starting number and ask them to write it down on a strip of paper. They add 25 to it. Then they add 25 to their answer. They continue, creating a series of 12 numbers. They should work together to check that they do this accurately. Do they notice a pattern? Can they predict what the twentieth number would be? What would happen if they added 12·5 every time? Try a few numbers to check.

ACTIVITY 2
Pairs

• *Repeatedly adding 11 to a given number*

Strips of paper

Each child chooses a starting number and writes it on their strip of paper. They add 11 to it. Then they add 11 to their answer. They continue creating a series of 12 numbers. They should work together to check that they do this accurately. Do they notice a pattern? Can they predict what the twentieth number would be?

ACTIVITY 3
Pairs

• *Finding palindromic numbers*

Explain to the children that to find a palindromic number (one that reads the same forwards as backwards) you write down the number, reverse its digits and add. Let the children do this until they obtain a palindromic number, for example, 68 + 86 = 154, 154 + 451 = 605, 605 + 506 = 1111. They then find all the palindromic numbers they can make using the numbers up to 30.

GROUP ACTIVITY 4
2–3 children

• *Exploring Pascal's triangle*

GROUP ACTIVITY 5
Pairs

• *Finding the digital root of a 4-digit number, working out a missing digit*

N40

Pascal patterns

Addition/subtraction
GROUP ACTIVITY 4

2–3 children

It is believed that this triangle of numbers was first invented by Jia Xian in China in the twelfth century.

Pascal made the triangle famous in the West in the seventeenth century.

Copy the triangle of numbers shown.

Extend it by adding another five rows on the bottom.

Each number is the sum of the two numbers above it.

Add the numbers in each row. What do you notice?

Add the numbers in the diagonal rows to find more patterns.

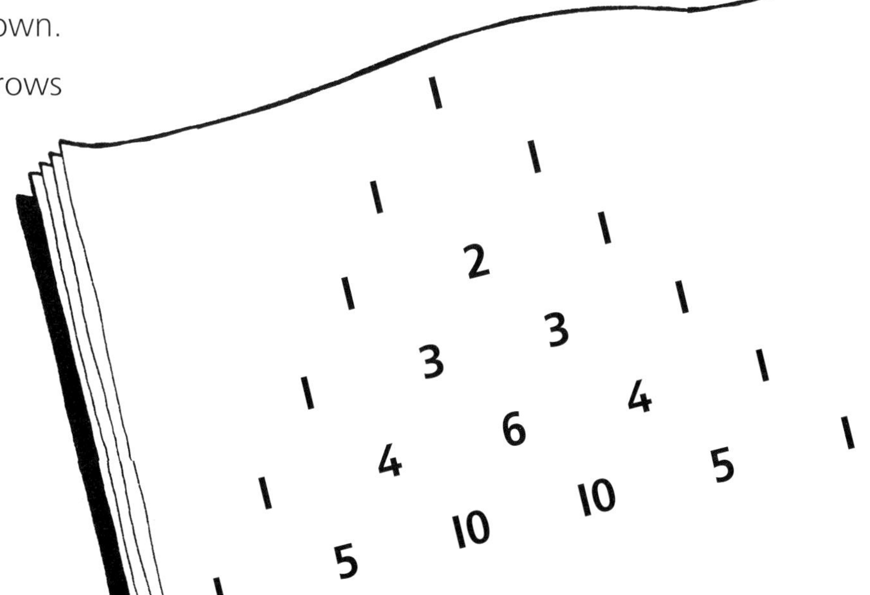

N40

Missing digit

Addition/subtraction
GROUP ACTIVITY 5

Pairs

Try this trick. One of you secretly write down any 4-digit number in which every digit is different, for example, 5132.

Find the digital root: 5 + 1 + 3 + 2 = 11, 1 + 1 = 2.

Cross out one of the original digits to get a new 3-digit number, for example, 512. Show this to your partner and tell them the digital root of your original number.

Your partner subtracts the digital root from the new 3-digit number. They find the digital root of this new number and subtract it from 9 to find the missing digit.

Try this again, swapping roles. Use ten different numbers to make sure it works.

Does it work with a 5-digit number?

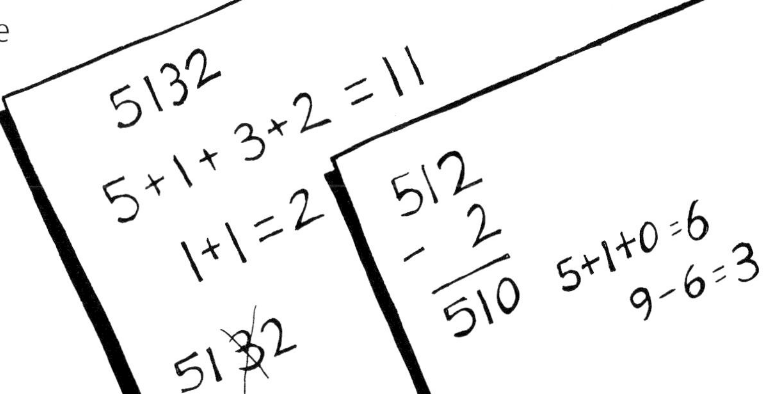

Abacus Ginn and Company 2001 Copying permitted for purchasing school only. This material is not copyright free.

N41 Properties of number

ACTIVITY 1
Whole class, in groups of 3

• *Estimating and calculating square roots*

Number cards (20 to 100) (PCMs 10 to 15), counters, a calculator

Shuffle the cards and place them in a pile face down. One child takes a card. The other two decide which two numbers the square root of that number will be between. For example, if the card number is 35, the square root will be between 5 and 6 (5 × 5 = 25 and 6 × 6 = 36). They also have to say which number the square root will be nearer to, i.e. it will be nearer to 6. The first child then works out the square root of the card number using the calculator. Was their answer correct? If so, they each take a counter. Repeat at least 12 times with the children taking different roles. Who has the most counters?

ACTIVITY 2
Pairs

• *Finding patterns in square numbers*

A calculator

The children write the square numbers up to 200. They find the digital root of each square number. They should work together to check that they do it accurately. Do they a pattern? Repeat for the cube numbers.

ACTIVITY 3
Pairs

• *Illustrating square numbers*

Squared paper, a calculator

Using the squared paper, the children cut out different-sized squares to illustrate square numbers up to 200. They write the number in each square that they cut out. They then use the squares to help them predict the square roots of the following numbers: 120, 180, 225. They can use a calculator to check how close the sizes of their squares are to the actual numbers.

ACTIVITY 4
Pairs

• *Estimating square roots*

Spreadsheet software

Prepare a spreadsheet file similar to the one shown. Use the 'If logical' functions **=IF(B3>0,IF(B3=SQRT(A3)," ","Try again")," ")** in Column C, **=IF(B3>0,IF(B3=SQRT(A3),B3," ")," ")** in Columns D and F, and **=IF(H3>0,IF(H3=D3*F3,"Well done"," ")," ")** in Column I.
In Column B the children estimate the square root of the number in Column A. If they are correct a multiplication will appear in columns D to G. The children enter the answer in Column H and check Column I to see if they are correct.

	A	B	C	D	E	F	G	H	I
1	Number	Square root		Square root	multiplied by		equals	Number	
2	2809	53		53	×	53	=	2809	Well done
3	729	27		27	×	27	=	729	Well done
4	1225	25			×		=	=	Well done
5	16		Try again		×		=	=	Well done

GROUP ACTIVITY 5
4 children

• *Estimating and calculating square roots*

A calculator, a dice (1 to 6), interlocking cubes

GROUP ACTIVITY 6
Pairs

• *Squaring and adding digits*

In between

4 children

Properties of number
GROUP ACTIVITY 5

N41

A calculator, a dice (1 to 6), interlocking cubes

Take turns to roll the dice. Choose a number from the column for the number you rolled in the table below.

Say between which two numbers the square root of your chosen number lies.
For example, if you choose 90, you might say that the square root is between 9 and 10.

The others use a calculator to check the exact square root.

Take a cube if you were correct.

Continue playing until all the numbers have been chosen.

Who has the most cubes?

Dice numbers					
1	2	3	4	5	6
90	300	220	180	39	0·5
24	1000	125	40	800	250
56	404	65	12	600	101
909	15	750	1700	5	20

Zapping

Pairs

Properties of number
GROUP ACTIVITY 6

N41

Write down a 2-digit number, for example, 24.

Square each of its digits: 4 and 16.

Add the two together: 4 + 16 = 20.

Repeat the process, squaring each digit and adding: 4 + 0 = 4.

Keep going until you are left with a 1-digit number.

Each try several 2-digit numbers. Numbers which reduce down to 1 are called 'happy numbers'. Can you find any?

24
$2^2 = 4$
$4^2 = 16$
$4 + 16 = 20$
$2^2 = 4$
$0^2 = 0$
$4 + 0 = 4$

Abacus Ginn and Company 2001 Copying permitted for purchasing school only. This material is not copyright free.

N42 Properties of number

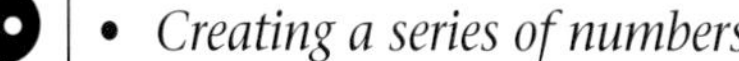

ACTIVITY 1
Whole class, in pairs

- *Creating a series of numbers*

A dice (1 to 6), interlocking cubes

Ask each child to write down a series of numbers. Give them a starting number, ask them to add a number to get the next number, and then add the same number to get the next number and so on. For example, they might start with 5 and add 3, creating the series 5, 8, 11, 14, 17, 20, 23, 26, 29, 32, 35, 38 ... They each write the first 12 numbers in their series. The children then roll the dice twice. They use the numbers to make two 2-digit numbers. For example, if they roll 4 and 2, they can make 42 and 24. If they roll the same number twice, they only make one number. If either of the numbers is in either of the pair's series of numbers, they may take a cube. The children roll the dice ten times. Who has the most cubes? Repeat, starting with a new number.

ACTIVITY 2
Pairs

- *Creating a series of numbers*

A dice (1 to 6)

Each child writes down a series of numbers. They choose a starting number, add a number to get the next number, then add these two numbers together to get the third number. They then add the second and third number together to get the fourth number and so on. For example, they might start with 3 and add 4, creating the series 3, 7, 10, 17, 27, 44, 71, 115 ... Each child writes the first 12 numbers in their series. The children swap number series and circle the even numbers in each series. They also draw a square around each multiple of 5 and a triangle around each multiple of 3. What do they notice?

ACTIVITY 3
Pairs

- *Exploring patterns of triangular numbers*

Spreadsheet software

Prepare a spreadsheet file similar to the one shown. Use the 'If logical' functions **=IF(C4>0,C3+C4," ")** in Column D and **=IF(E3>0,IF(E3=B3*B3,"Well done","Try again")," ")** in Column E. As the children enter the sequence of triangular numbers in Column C, the sequence of square numbers will be generated in Column D. They check Column E to see if their answers are correct and try to find how the triangular and square numbers are connected.

	A	B	C	D	E
1	Term		Triangular numbers	Square numbers	
2	1	st	1	1	Well done
3	2	nd	3	4	Well done
4	3	rd	6	9	Well done
5	4	th	10	16	Well done
6	5	th	16	26	Try again
7	6	th			
8	7	th			
9	8	th			
10	9	th			
11	10	th			

GROUP ACTIVITY 4
Pairs

- *Investigating numbers in a circle*

GROUP ACTIVITY 5
Pairs

- *Investigating patterns of numbers*

Day wheel

Pairs

Choose a number, for example, 17.

Look at the numbered circle.

Starting at 1, count to 17 around the sections of the circle. You should finish on 3.

Try some other numbers. Where does 57 end up? Where does 36 end up? What about 93? Try some other large numbers.

Can you find a rule which means that you do not have to count around the circle each time? Try dividing your numbers by 7. What do you notice?

Look at the circle labelled with the days of the week. If today is Friday, what day will it be in 32 days time? Use your rule to help you find the answer quickly.

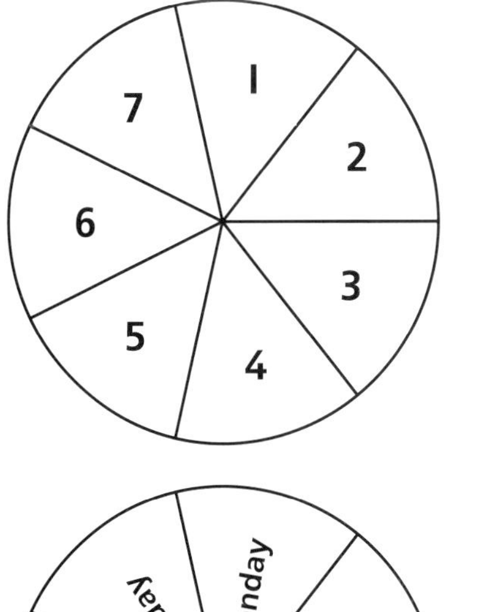

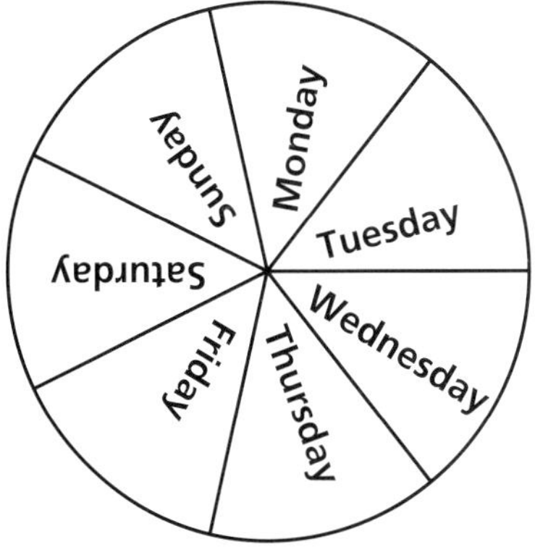

Number triangle

Pairs

Copy these additions. They are the first two lines of a triangle. Can you write the third and fourth lines, following the pattern, and see if they work?

What about the fifth and sixth lines?

Try out some other arrangements of 1-digit numbers. Try 2 + 3 + 7 = 1 + …

1 + 2 = 3

4 + 5 + 6 = 7 + 8

Abacus Ginn and Company 2001 Copying permitted for purchasing school only. This material is not copyright free.

N43 Properties of number

ACTIVITY 1
Whole class, in pairs

• *Reducing odd and even numbers and investigating the patterns they produce*

Two large strips of paper

Ask the children to write the odd numbers up to 50 in a line along their strip of paper. They work together to write the reduced number of each number. What patterns do they find? Now they think about the even numbers. Will these reduced numbers generate the same patterns? The children take the second strip of paper and write the even numbers up to 50. Working together, they write the reduced numbers. What patterns do they notice?

ACTIVITY 2
Pairs

• *Creating series of numbers and investigating their patterns*

Large strips of paper

On one of the strips of paper, the children write two starting numbers, one even and one odd, for a number series. They work together to generate a series by adding together the two starting numbers to get the third number, then adding the second and the third numbers to get the fourth number and so on. They write the first ten numbers of their series. They now circle every even number. What do they notice? Using another strip of paper, the children repeat the activity, starting this time with an odd and an even number. Do they obtain the same pattern of odds and evens? They do the activity again, this time with two even numbers. What is the pattern now? Finally they start with two odd numbers. What is the pattern this time?

ACTIVITY 3
4 children

• *Adding odd and even numbers*

Two large strips of paper, post-it notes, a calculator

The children write the odd numbers up to 50 in a line along their strip of paper. Two children work together to calculate the sum of the numbers. The other two children add the numbers using a calculator. They compare their answers. If their answers do not agree, they find a third way of adding them using both a calculator and some mental arithmetic, for instance, by adding the numbers in batches of six. When they have agreed a total, they write it on a post-it note. The children then write the even numbers to 50 along the second strip of paper and go through the same procedure.

GROUP ACTIVITY 4
Pairs

• *Creating patterns with odd numbers*

A calculator

GROUP ACTIVITY 5
Pairs

• *Creating odd and even numbers*

A dice (1 to 6)

N43

Arranging oddness

Pairs

Properties of number
GROUP ACTIVITY 4

A calculator

Copy this triangle.

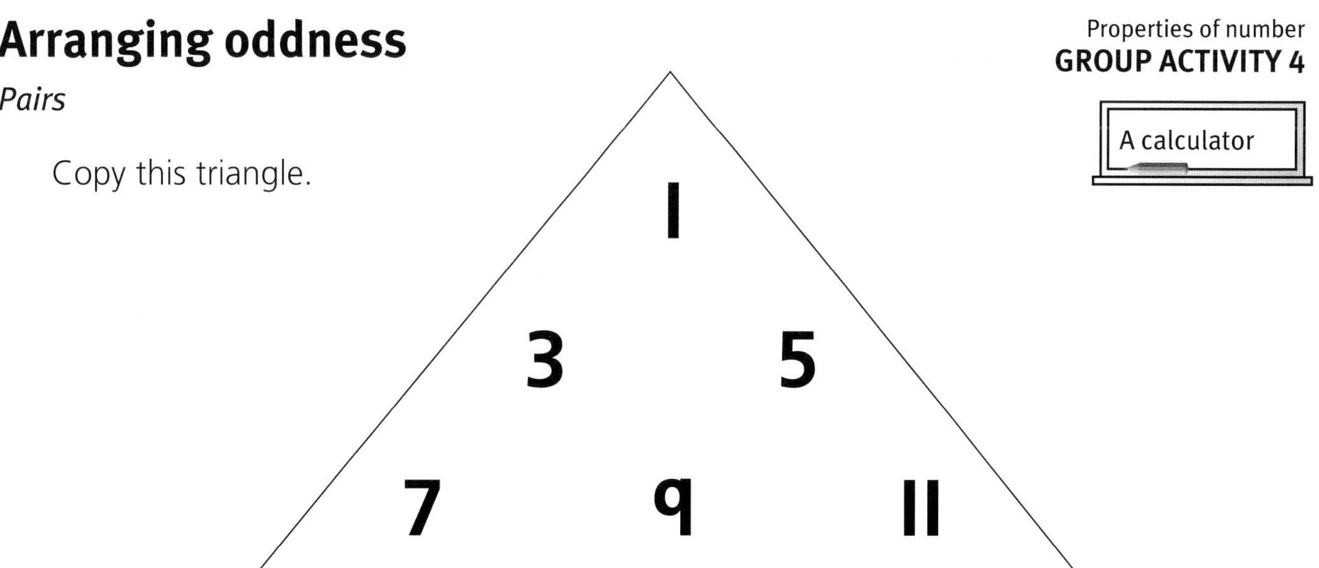

Work out how the triangle is formed.

Can you write the next four lines, following the pattern?

Add up the numbers in each line, writing the total beside it in a circle.

What do you notice about the totals?

Can you predict the total for the next five rows? Use a calculator to help you.

N43

Game luck

Pairs

Properties of number
GROUP ACTIVITY 5

A dice (1 to 6)

Take turns to play. Choose a balloon number.

Roll the dice. Multiply the balloon number by the number shown on the dice.

Score 11 for an odd number and 4 for an even number. Write down your score.

After seven rounds, who has the most points?

Abacus Ginn and Company 2001 Copying permitted for purchasing school only. This material is not copyright free.

M1 Length

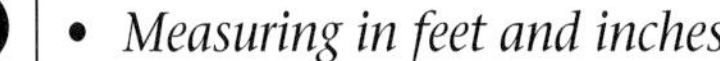

ACTIVITY 1
Whole class, in pairs

- *Measuring in feet and inches*
- *Converting feet and inches to metres and centimetres*

A ruler (marked in inches), string, scissors

Ask the children to take turns to cut a piece of string the same length as their partner's height. Each child measures the piece of string for their own height in feet and inches. Discuss together how to convert the measurements into metres and centimetres. The children then write their heights in imperial and metric measurements. Discuss the different heights. Who is tallest? Are any children the same height?

ACTIVITY 2
3 children

- *Drawing and measuring lines in inches*

A ruler (marked in inches), a dice (1 to 6)

One child throws the dice. The next child draws a line that number of inches long. The third child measures the line to check that it is the correct number of inches. They continue, changing roles each time, to build up a pattern by drawing the lines to criss-cross over each other.

ACTIVITY 3
Pairs

- *Measuring in millimetres, centimetres and metres*

A ruler (marked in cm and mm), a metre rule

In the classroom the children find three items to measure in millimetres. They draw each one and then measure it, checking that they both agree on the measurements. The children take turns to write the measurements beside the drawings. They repeat, finding three items to draw and measure using centimetres, then metres.

ACTIVITY 4
3 children

- *Converting from miles to kilometres*

A distance chart (giving distances between towns in miles)

The first child chooses one town and the second child chooses another. They each run a finger along the row or column from their town until the two fingers meet. The third child reads off the distance between the towns in miles and writes it down. The children work together to calculate the approximate distance between the towns in kilometres and write it next to the number of miles. They repeat the activity at least 12 times, sharing the roles.

GROUP ACTIVITY 5
3–4 children

- *Measuring in inches*
- *Converting inches to centimetres*

A ruler (marked in inches), coloured pencils

GROUP ACTIVITY 6
2–3 children

- *Converting miles to kilometres*

A large road map

Line lengths

Length
GROUP ACTIVITY 5

3–4 children

A ruler (marked in inches), coloured pencils

Take turns to choose a line and measure its length in inches.

Write the length at one end of the line.

Choose a coloured pencil and colour in one space between the lines in your colour.

Continue until all the lines have been measured.

Work together to convert all the measurements into centimetres.

How far?

Length
GROUP ACTIVITY 6

2–3 children

A large road map

Choose a starting point on the map.

Imagine you are on a cycling holiday.

Plan a route for one day's cycling, and check how many miles it is.

Write the number of miles and convert it into kilometres.

Plan a different route for each day for a week.

How many miles in total? How many kilometres in total?

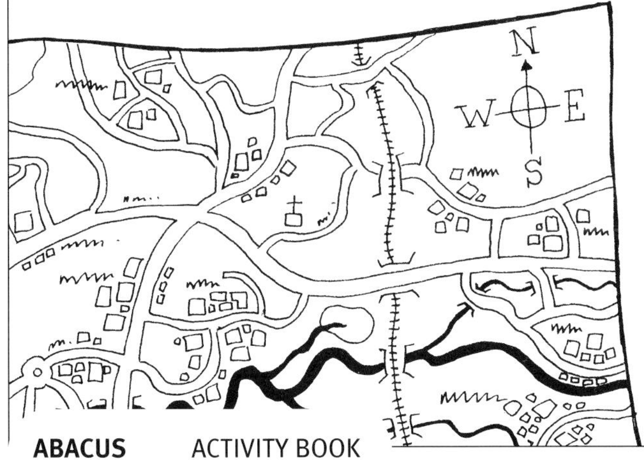

Day 1

15 miles → 24 Km

Abacus Ginn and Company 2001 Copying permitted for purchasing school only. This material is not copyright free.

M2 Weight

ACTIVITY 1
Whole class, in pairs

- *Estimating and measuring the weight of objects*
- *Using metric and imperial measurements of weight*

A balance, weights (4 oz, 100 g), interlocking cubes

Choose the 4 ounce weight. Ask the children to estimate how many cubes will be needed to balance 4 ounces. All the children write down their estimate. Place the 4 ounce weight in one side of the balance and add cubes to the other side until it is level. How many were needed? Now repeat the activity for the 100 g weight. How many cubes weighed 100 g? How close were the children's estimates? Repeat this activity with pound weights.

ACTIVITY 2
Pairs

- *Estimating and measuring the weight of objects*
- *Using metric and imperial measurements of weight*

A balance, weights (in kilograms, grams, pounds, ounces), a variety of objects to weigh

One child chooses an object. They estimate its weight in ounces (and pounds if it is heavy). They write down their estimate. The other child then weighs the object in pounds and ounces and writes down the weight. The children then compare the actual weight with the estimate. How close were they? The first child now estimates the weight in kilograms and grams. The second child reweighs the object using kilograms and grams. How close was the first child's guess? The children repeat the activity, swapping roles.

ACTIVITY 3
Pairs

- *Converting units of weight*

Guinness Book of Records

Together, the children look through the *Guinness Book of Records* searching for records which are based on weight. They write down several records and then work together to convert each one into a different unit, for example, they could convert grams into ounces or kilograms into tonnes. They then rewrite each selected record using a different unit of weight.

ACTIVITY 4
Pairs

- *Converting kilograms to pounds, and vice versa*

Bathroom scales (calibrated in kilograms)

The children weigh each other using the scales and write down their weights. They work together to convert them into pounds. They then find something heavy in the classroom, perhaps a large book and guess its weight in pounds. They weigh the item in kilograms and convert it into pounds. How close was their guess? They repeat the activity with another heavy object.

GROUP ACTIVITY 5
3–4 children

- *Estimating the weight of objects*

Interlocking cubes

GROUP ACTIVITY 6
4–5 children

- *Converting from imperial to metric units of weight*

Margarine, sugar, syrup, cocoa, digestive biscuits, chocolate, a saucepan, a baking tray

Match the weight

3–4 children

Interlocking cubes

Cover each rectangle with a cube.

Take turns to play.

Uncover a rectangle by removing the cube. Read the weight.

Think about something which weighs that amount.

Think carefully then say it out loud.

The others must agree that this is a good approximation. If so, keep the cube.

Play until all the weights are uncovered.

Recover the squares and play again, finding different examples.

4 oz	8 oz	1 lb	1 tonne	100 kg
100 g	250 g	1 kg	5 kg	100 lb

Chocolate biscuit cake

4–5 children

Margarine, sugar, syrup, cocoa, digestive biscuits, chocolate, a saucepan, a baking tray

Use the following quantities of the ingredients to make some chocolate biscuit cake.

Convert each quantity into its metric equivalent and write out the recipe again.

Ingredients

4 oz hard margarine or butter
2 oz sugar
3 oz golden syrup
4 oz cocoa
10 oz digestive biscuits

- Melt all the ingredients except the biscuits and stir until bubbling.
- Crush the biscuits and stir them into the chocolate mixture.
- Press into a baking tray and cool in the fridge for two hours.

Abacus Ginn and Company 2001 Copying permitted for purchasing school only. This material is not copyright free.

M3 Capacity

ACTIVITY 1
Whole class, in pairs

- *Estimating the capacity of containers*

Plastic bottles (of different capacities, with contents labels)

Let the children study the different bottles and look at the amounts they hold. Ask some of the children to arrange the bottles in order from smallest to largest. All the children write down their capacities in order. They then estimate the capacity of each bottle in pints and write them down.

ACTIVITY 2
Pairs

- *Using centilitres and millilitres*

An empty 75 cl bottle with the label removed, a measuring cylinder marked in 10 ml divisions, water, a dice (1 to 6)

The children take turns to throw the dice and measure that number of centilitres of water in the measuring cylinder. They write down the number of centilitres needed and convert this to millilitres to find how much water to measure out. They pour the water into the bottle and repeat, keeping a note of how much water they use each time. How many times do they need to throw the dice to fill the bottle? They work out the capacity of the bottle in centilitres and millilitres by adding up the amount of water poured in each time?

ACTIVITY 3
Pairs

- *Using pints and litres*

A supermarket or grocery catalogue

The children look through the catalogue items and find as many items as they can which they think would be measured in litres and in pints. They also think of other liquids which might be measured in either pints or litres. Using two headings, 'Pints' and 'Litres', the children list as many liquids as they can, saying how many litres or pints of each people would usually buy at any one time.

GROUP ACTIVITY 4
Pairs

- *Measuring the capacity of containers*

A tablespoon, lentils, a variety of containers

GROUP ACTIVITY 5
Pairs

- *Comparing pints and millilitres*

Two 5 ml medicine spoons, lentils or rice, a pint container

Which holds more?

Pairs

Capacity
GROUP ACTIVITY 4

A tablespoon, lentils, a variety of containers

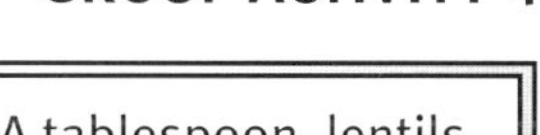

Choose a container each.

Fill them with lentils using the spoon. How many spoonfuls does each hold? Write down the numbers.

Empty both containers and swap.

Repeat with your new container.

Compare your answers.

Which container holds more?
How much more?

How many millilitres?

Pairs

Capacity
GROUP ACTIVITY 5

Two 5 ml medicine spoons, lentils or rice, a pint container

Take turns to pour spoonfuls of lentils into the pint container.

Keep a running total of the number of millilitres: 5 ml for each spoonful.

How many millilitres does the container hold?

Repeat once more.

Compare the two answers.

Abacus Ginn and Company 2001 Copying permitted for purchasing school only. This material is not copyright free.

M4 Area

ACTIVITY 1
Whole class, in pairs

- *Calculating the area of rectangles*

Cm-squared paper

The children draw shapes which can be split into rectangles, e.g. L shapes, T shapes. They calculate the area of each drawing and write it beside the drawing. They swap drawings and check each other's calculations. Are the areas correct? The children repeat the activity for different shapes.

ACTIVITY 2
Pairs

- *Calculating the area of rectangles*

Two 30 cm × 30 cm squares drawn on cm-squared paper, a dice (1 to 6)

Taking turns, each child throws the dice twice. On their 30 × 30 square they draw a rectangle with the dimensions given by the dice throw, for example, if they throw a 6 and a 4, they draw a 6 cm × 4 cm rectangle in one corner of their 30 × 30 square. They shade the rectangle and write the area i.e. 24 cm^2. When each child can colour no more rectangles in their square they each add up the total of all the areas of all their rectangles. The child whose total is closest to 900 square cm wins.

ACTIVITY 3
3–4 children

- *Calculating the area of rectangles*

Cm-squared paper, number cards (2 to 10) (PCM 9)

The children take two cards each. They draw a rectangle with dimensions given by the cards, for example, they take a 3 and an 8 and draw a rectangle 3 cm by 8 cm. They calculate its area. They then add or subtract a section to make their rectangle into an L shape with an area of 30 square cm. The children check each other's L shapes. Do they all have an area of 30 cm^2? They reshuffle the cards and play again.

GROUP ACTIVITY 4
3 children

- *Calculating the area of rectangles*

GROUP ACTIVITY 5
3–4 children

- *Calculating the area of rectangles*

Interlocking cubes, cm-squared paper, a dice (1 to 6)

Guess the area

Area
GROUP ACTIVITY 4

M4

3 children

Choose a shape each. Do not tell the others which one you have chosen.

Work out the area of your shape and write it down.

Show the areas to each other. In turn, work out which shape you each chose.

Repeat, choosing another shape each.

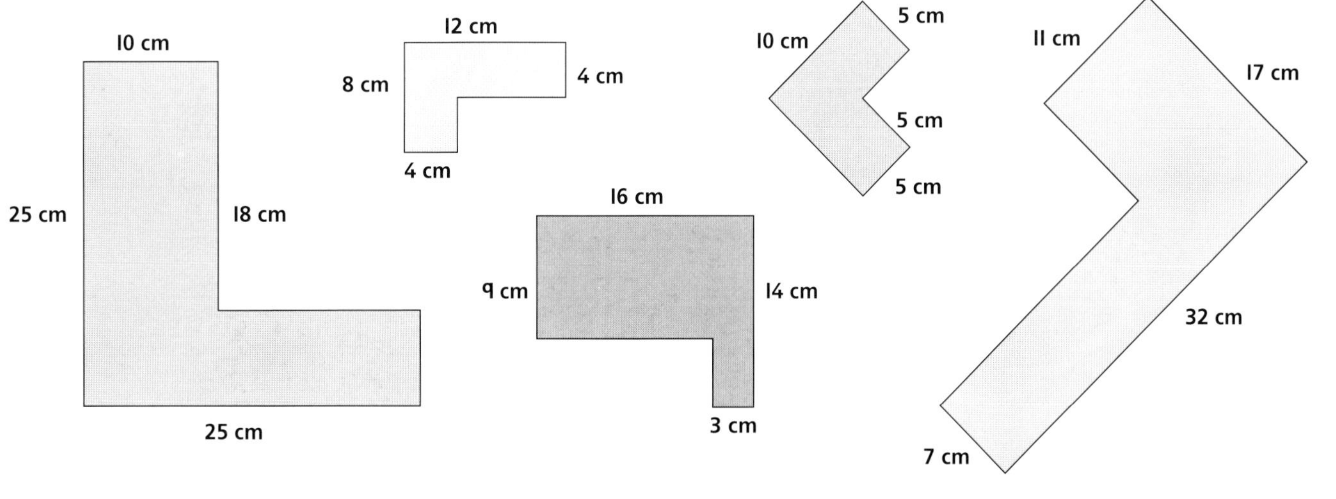

Target area

Area
GROUP ACTIVITY 5

M4

3–4 children

Interlocking cubes, cm-squared paper, a dice (1 to 6)

Your aim is to draw an L shape.

Take turns to throw the dice.

Draw a line the same number of squares long as the dice throw.

Continue until you have completed an L shape. (You will need to throw the dice four times). For example, you could throw a 6 then a 3 then a 4 then a 1.

Find the area of your L shape.

The child whose area is closest to 30 takes a cube.

Continue until someone has collected five cubes.

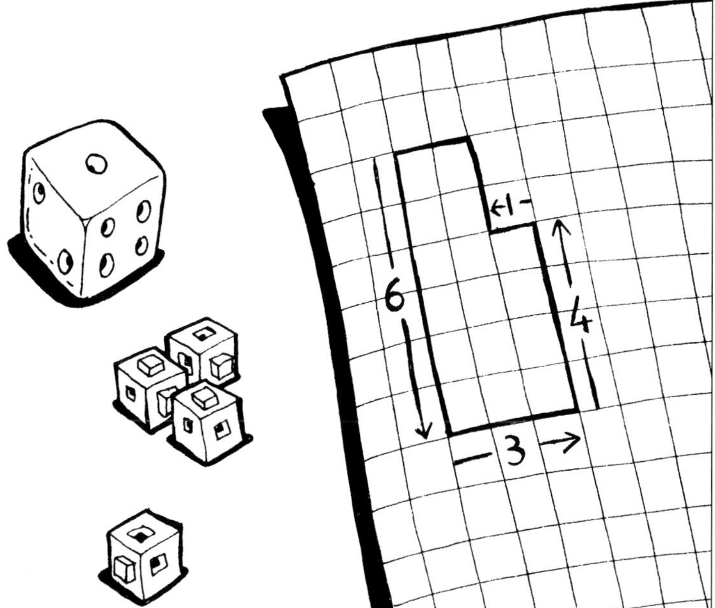

Abacus Ginn and Company 2001 Copying permitted for purchasing school only. This material is not copyright free.

M5 Area

ACTIVITY 1
Whole class, in pairs

- *Understanding the area of a triangle as half the area of a rectangle*

Cm-squared paper, scissors

Each child draws a rectangle, cuts it out and writes the area of the rectangle. They fold the rectangle in half along its diagonal and cut out the two triangles. They pass one of the triangles to their partner. They each calculate the area of the triangle and write it on the back. They then compare the area of each triangle with the area of the rectangle it came from. The children repeat the activity, starting with different rectangles.

ACTIVITY 2
3 children

- *Calculating the area of right-angled triangles*

Cm-squared paper, a ruler

One child draws a right-angled triangle on squared paper. The second child measures one side (not the longest) and writes it down. The third child measures another side (not the longest) and writes that down. The first child calculates the area of the triangle. The others check to see if they agree and they write the area in the middle of the triangle. The children then count the squares to see if their calculation was correct. They repeat the activity, swapping roles, until they have drawn at least six triangles.

ACTIVITY 3
Pairs

- *Finding the length of the side of a triangle*

Cm-squared paper, a ruler

Each child draws a right-angled triangle on squared paper. They calculate its area. They do not show their partner – but they tell them the area and also the length of one side. The children then reconstruct the triangle that their partner drew and compare the results. They repeat the activity starting with a new triangle each.

GROUP ACTIVITY 4
2–3 children

- *Finding the area of 2-d shapes*

Regular 2-d shapes, a ruler, a calculator

GROUP ACTIVITY 5
2–3 children

- *Calculating the area of rectangles and triangles*

Large cardboard rectangles, rulers, scissors, a calculator

Shape sizes

Area
GROUP ACTIVITY 4

2–3 children

Regular 2-d shapes, a ruler, a calculator

Working together, draw around one of the shapes.

Decide the best way to split that shape into rectangles and right-angled triangles to find its area.

Take the measurements you need and write them on your drawing.

Use the calculator to help you find the area of each triangle and rectangle.

Add them together to find the total area of the shape.

Do the same for each shape in turn.

Place them in order, from smallest area to largest.

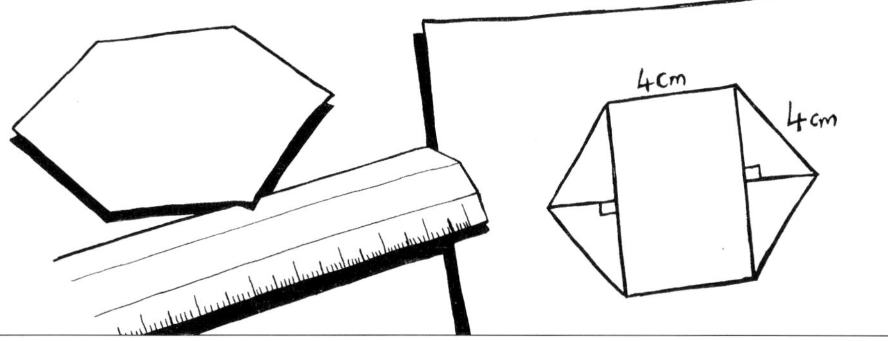

Jigsaws

Area
GROUP ACTIVITY 5

2–3 children

Large cardboard rectangles, rulers, scissors, a calculator

Choose a rectangle each. Measure its sides, find its area and write it down.

Divide the rectangle up, using three straight lines. Cut the rectangle into smaller pieces along the lines.

Each pass the smaller pieces to the person on your left.

Find the area of each smaller piece (dividing them into rectangles and right-angled triangles if you need to). Use a calculator to help.

Find the total area of the smaller pieces. Compare it with the area of the original rectangle.

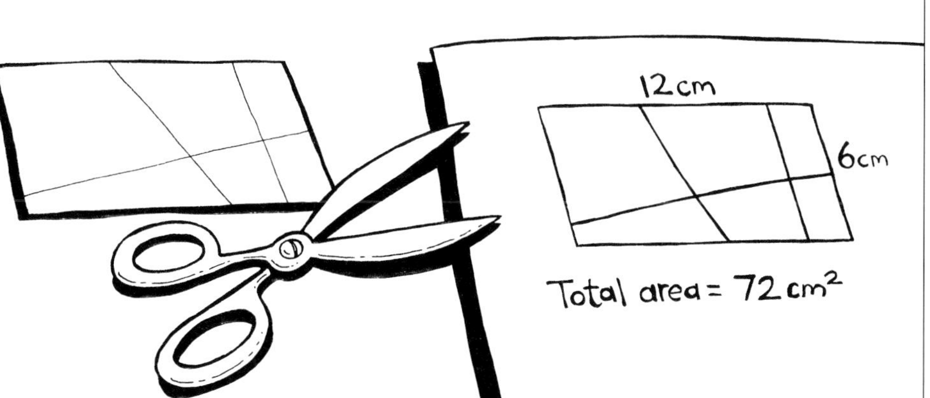

Abacus Ginn and Company 2001 Copying permitted for purchasing school only. This material is not copyright free.

M6 Perimeter

ACTIVITY 1
Whole class, in groups of 3

- *Calculating and measuring the perimeters of rectangles*

Books, string, a ruler

One child chooses a book. The second child measures one side and the third child measures the other side. The first child uses their measurements to calculate the perimeter. They agree the answer and write it down. The children then wrap a piece of string around the edge of the book and cut it off so it is exactly the length of the perimeter. They measure the string to check their calculation. How accurate were they? The children choose other books and repeat the activity several times.

ACTIVITY 2 
2–3 children

- *Drawing polygons with given perimeters*

Cm-squared paper, rulers

Each child draws a different shape on the squared paper. The perimeter must be exactly 60 cm. They construct as many different shapes as they can, all with a perimeter of 60 cm.

ACTIVITY 3 
Pairs

- *Drawing polygons with given perimeters*

Cm-squared paper, rulers

How many different polygons can the children draw which have a perimeter of 36 cm? All the shapes must have sides which are an exact number of centimetres. How many of these are regular shapes?

GROUP ACTIVITY 4 
Pairs

- *Finding the perimeter of regular shapes*

Cm-squared paper, rulers

GROUP ACTIVITY 5 
Pairs

- *Finding the perimeter of irregular shapes*

Cm-squared paper, rulers

Perimeter patterns

Perimeter
GROUP ACTIVITY 4

Pairs

Cm-squared paper, rulers

Find the perimeter of each of these shapes.

Draw the next shape in the series.

What is its perimeter?

Draw the next shape.

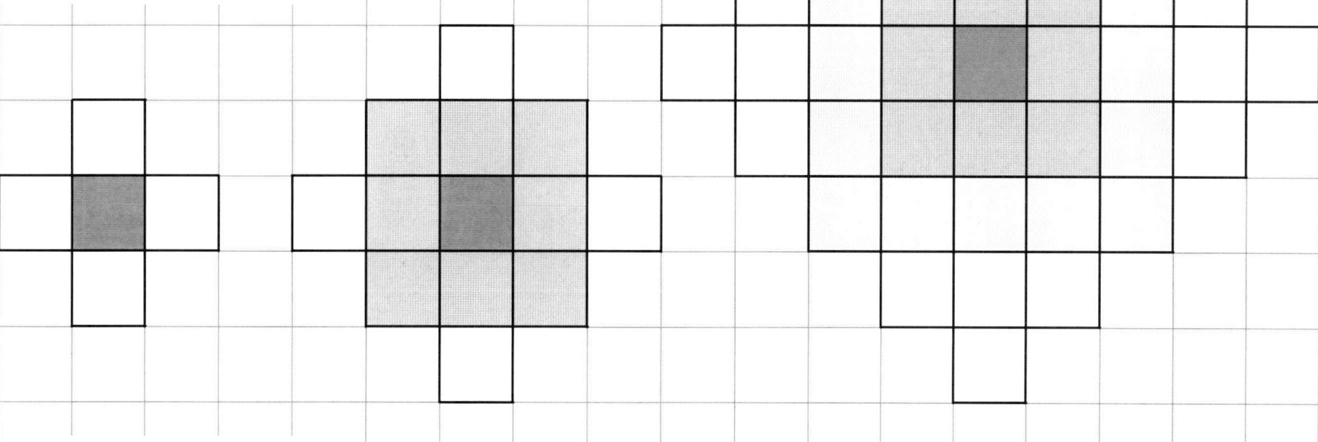

Floor plan

Perimeter
GROUP ACTIVITY 5

Pairs

Cm-squared paper, rulers

Draw a floor plan of your classroom.

Work out its perimeter.

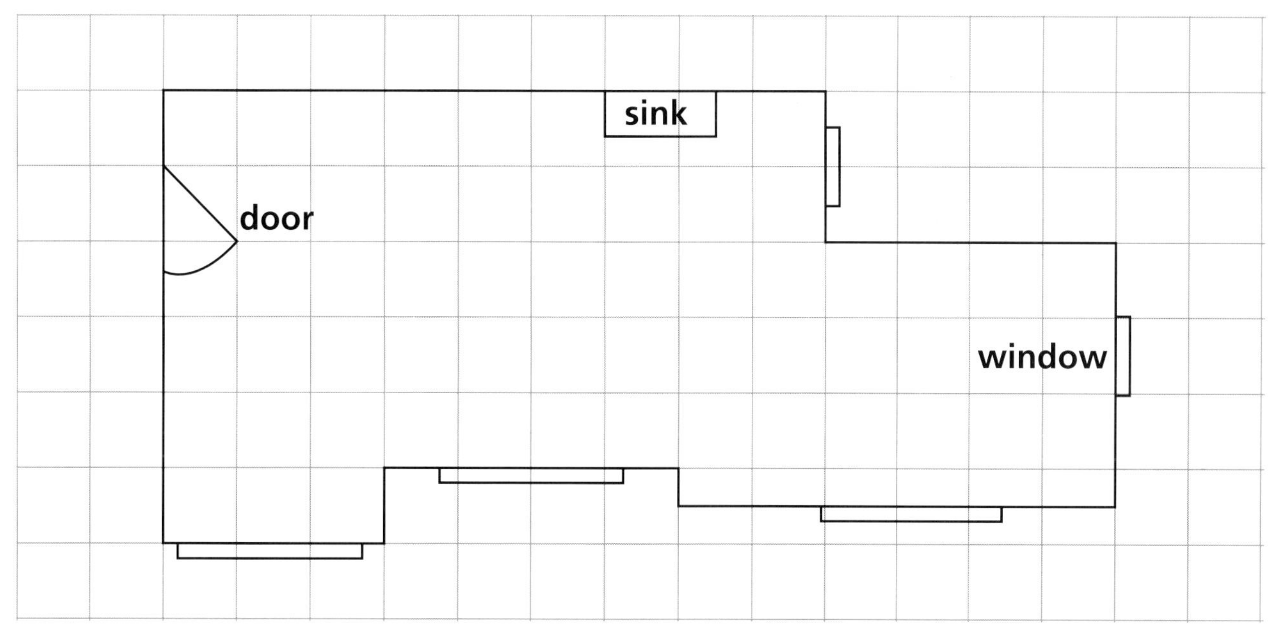

Abacus Ginn and Company 2001 Copying permitted for purchasing school only. This material is not copyright free.

S1 Angle

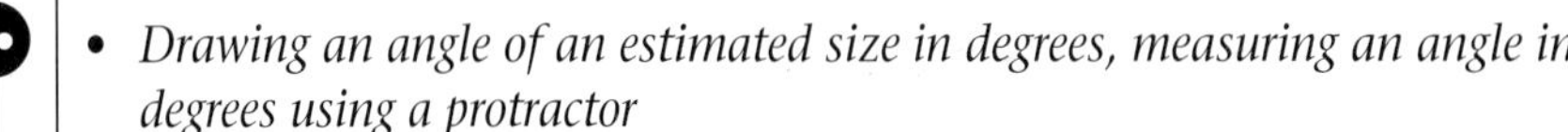

ACTIVITY 1
Whole class, in pairs

- *Drawing an angle of an estimated size in degrees, measuring an angle in degrees using a protractor*

Number cards (10 to 90) (PCMs 9 to 15), protractors, rulers

Shuffle the cards and place them face down in a pile. One child turns over the top card, e.g. 38. Each child draws an angle that they estimate to be 38°. They measure each other's angles using a protractor. Whose estimate is the closest? Repeat several times.

ACTIVITY 2
Pairs

- *Estimating the size of an angle in degrees, measuring an angle in degrees*

Two different coloured paper circles, a protractor, scissors

The children make the paper circles intersect by cutting a straight line from the circumference to the centre of each circle, then joining the two circles. One child chooses an angle at random, e.g. 58°. The other child rotates the circles until the angle of overlap is at an estimated 58°. They measure the angle created with the protractor. They calculate the difference between the estimated size and the target – this is the score. Repeat several times with the children swapping roles. The winner of each round is the child with the lower score.

ACTIVITY 3
4 children

- *Drawing an angle of a given size, using the protractor*

Place-value cards (T, U) (PCMs 1 to 4), a protractor, a ruler

The cards are shuffled separately and one of each type is turned over to create a 2-digit number, e.g. 56. Each child draws an angle of 56°. They check each other's angle and then write the size of the angles. Repeat for five different angles.

ACTIVITY 4
Pairs

- *Calculating the angles at a point, given one angle, measuring an angle in degrees using a protractor*

A ruler, a protractor

One child draws two long intersecting lines. The other child accurately measures the size of one of the angles using a protractor. The first child then calculates the size of each of the other three angles. They check the angles with a protractor. Are the angle sizes correct? Repeat, with the children swapping roles.

GROUP ACTIVITY 5
Pairs

- *Calculating angles at a point, given other angles*

Find the angles

Pairs

Write the angles inside each triangle.

Write the angle total for each triangle.

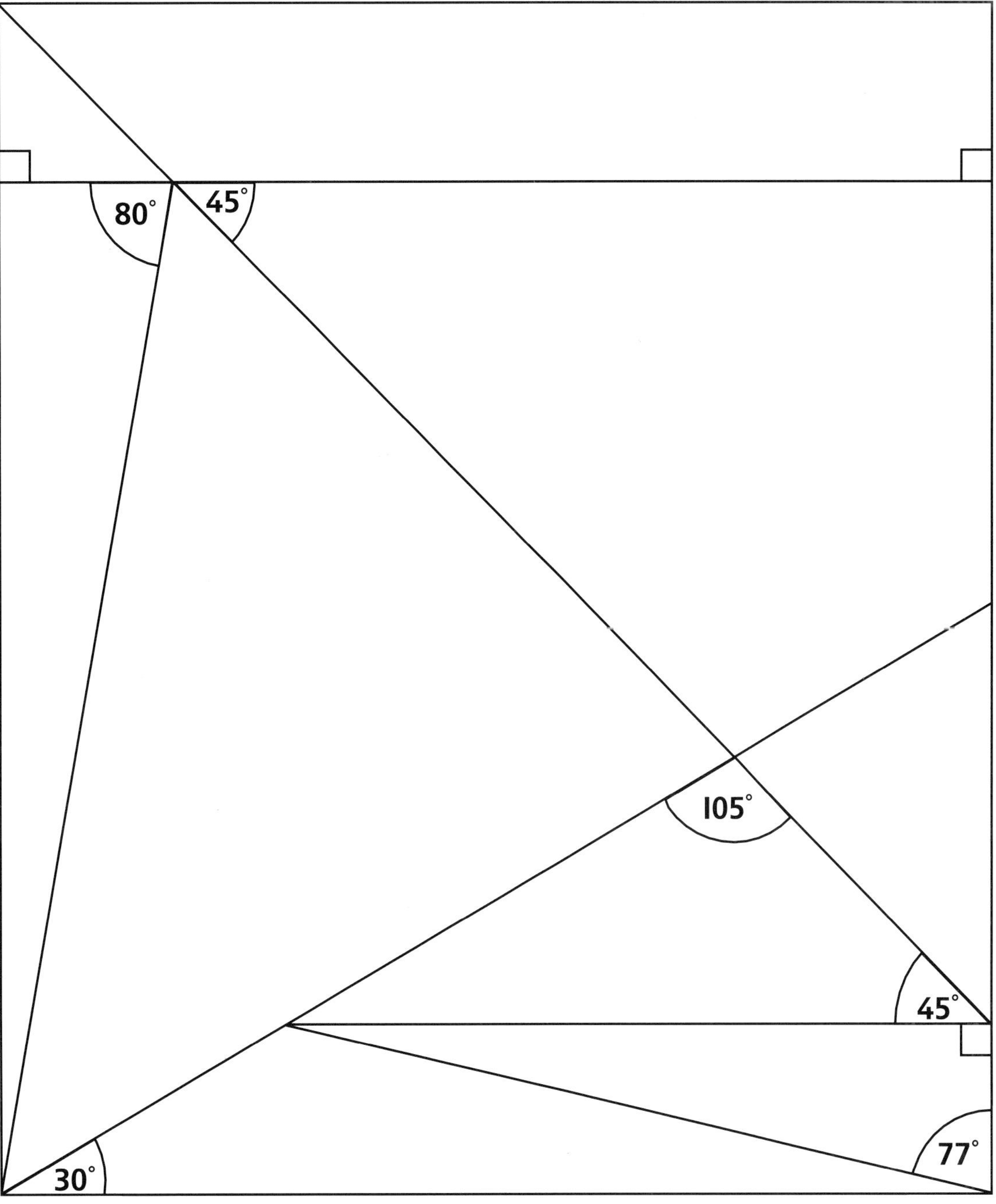

Abacus Ginn and Company 2001 Copying permitted for purchasing school only. This material is not copyright free.

S2 Angle

ACTIVITY 1
Whole class, in pairs

- *Measuring the angles of a triangle, recognising the angle sum of a triangle*

Rulers, protractors, squared paper, glue

Each child draws a large right-angled triangle on the squared paper and gives it to their partner. They measure one of the angles (not the right angle) and write the size in the angle. They pass on the triangles again and they each measure the remaining angle and write in the size. The children add up the angles for each triangle. If they do not total 180°, they work together to find the mistake. They each stick one of the triangles in their exercise book. Repeat.

ACTIVITY 2
Pairs

- *Measuring the angles of a triangle, recognising the angle sum of a triangle*

Several different cardboard rectangles, a protractor, a ruler, scissors, glue

Each child takes a rectangle, draws a diagonal across it and cuts it into two triangles. They each take one triangle. Which angle do they already know? They each measure one of the other angles in the triangle and write in the size. They exchange triangles and measure the remaining angle. Together, they add up the angles in each triangle. If they both total 180°, they stick an example of both triangles in their exercise book. Repeat with two different rectangles.

ACTIVITY 3
2–3 children

- *Measuring the angles of a triangle, recognising the angle sum of a triangle*

Several paper rectangles, protractors, a ruler, glue

The children each choose a rectangle and fold it in half. They draw a line from the mid-point of the top side to the bottom left-hand corner. They then do the same for the right-hand corner. They cut out their triangles and then compare them. All the triangles should be isosceles. Each child measures the angles of their triangle and writes in the sizes. They exchange triangles and check each other's measurements by adding the angles to check that their total is 180°. They stick the triangles in their exercise book. Repeat.

ACTIVITY 4
3–4 children

- *Exploring the angle sum of different polygons, measuring an angle in degrees using the protractor*

Rulers, protractors, large sheets of paper

Each child draws a large rectangle, then measures and marks each angle. They pass them around for checking. Each child finds the total of the four angles and these are then checked by the rest of the group. Do they notice anything about the totals? Repeat for a large pentagon, hexagon, etc.

GROUP ACTIVITY 5

Pairs

- *Measuring the angles of different right-angled triangles*

Squared paper, a protractor, a ruler

GROUP ACTIVITY 6
2–3 children

- *Drawing isosceles triangles of given angles*

Squared paper, protractors, rulers, glue

Right-angled triangles

Angle
GROUP ACTIVITY 5

Pairs

Squared paper, a protractor, a ruler

Draw a right-angled triangle with sides 3 cm and 3 cm.

Measure the other two angles.

Draw a right-angled triangle with sides 3 cm and 4 cm.

Measure the other two angles.

Draw a right-angled triangle with sides 3 cm and 5 cm.

Measure the other two angles.

Together, explore different right-angled triangles and the different angles.

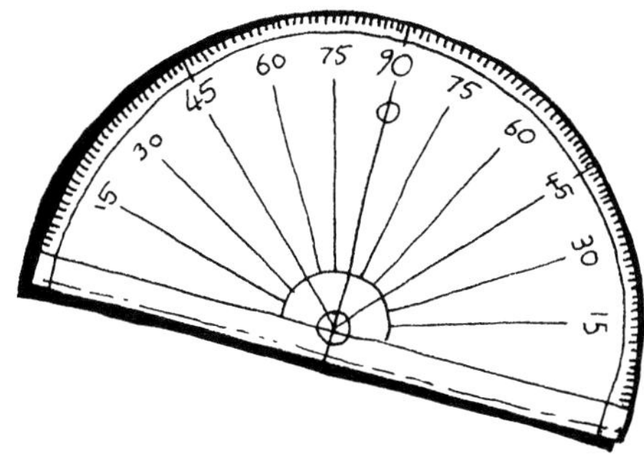

45°

45°

Isosceles triangles

Angle
GROUP ACTIVITY 6

2–3 children

Squared paper, protractors, rulers, glue

Each draw an equilateral triangle (with angles of 60°, 60° and 60°).

Compare your drawings and check that the triangles are identical.

Next, draw an isosceles triangle with angles 50°, 50° and 80°.

Compare your drawings again.

Each draw an isosceles triangle with angles of 40°, 40° and 100°.

Carry on, drawing a series of isosceles triangles, with smaller and smaller equal angles.

What do you notice about the shapes?

When you have finished, cut out the triangles and stick them onto a piece of paper to make a pattern.

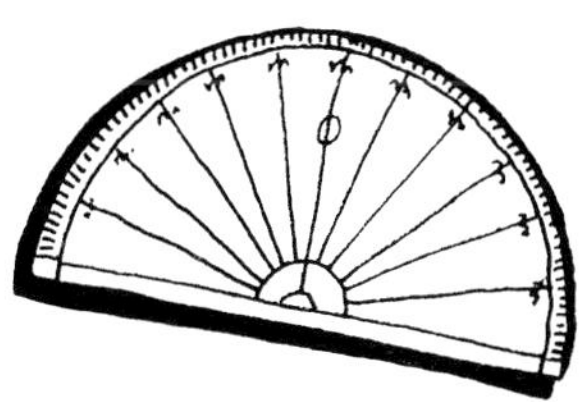

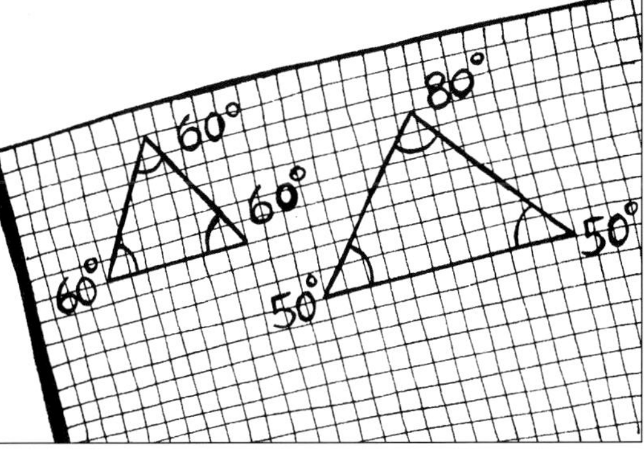

Abacus Ginn and Company 2001 Copying permitted for purchasing school only. This material is not copyright free.

S3 Coordinates

ACTIVITY 1
Whole class, in pairs

• *Plotting and reading points in four quadrants on a coordinate grid, recognising and naming quadrilaterals*

Coordinate grid (PCM 21), coloured pens or pencils

One child secretly draws a type of quadrilateral on their grid. They say the coordinates of the vertices. The other child plots them on their grid, joins the vertices to create the quadrilateral, and names it. They compare the two drawings. Repeat several times, sharing the roles and using a different colour for each shape, so that several shapes can be drawn on the same grid.

ACTIVITY 2
Pairs

• *Plotting and reading points in four quadrants on a coordinate grid*

Coordinate grid (PCM 21), counters

One child gives the coordinates of a point on the grid, e.g. negative two, five. The other child places a counter on the matching point on their grid, checking that it is correct by repeating the coordinates. The children name ten points each, swapping roles.

ACTIVITY 3
2–3 children

• *Describing a path on a grid using coordinates and compass directions*

Coordinate grid (PCM 21)

The children draw a polygon on their grid, using coordinates at its vertices. They then write a description of how to draw the polygon, based on compass directions, assuming that North is vertically up the page, e.g. 'Start at ($^{-}3$, 4), travel five squares north east, then turn to face west ...'. The children test their descriptions on each other, trying to draw their partner's polygon.

ACTIVITY 4
2 pairs

• *Plotting and reading points in four quadrants on a coordinate grid*

Two sets of number cards ($^{-}5$ to 5) (PCMs 8, 9), coordinate grid (PCM 21), counters

Shuffle each set of cards and place them in two piles face down, one pile for the horizontal coordinate, one for the vertical coordinate. Each pair takes turns to turn over one card from each pile and plots that point on the grid with a counter. The winning pair is the first to have three counters in a straight line. Play again several times.

GROUP ACTIVITY 5
3–4 children

• *Plotting and reading points in four quadrants on a coordinate grid*

Coordinate grid (PCM 21), three sets of number cards ($^{-}5$ to 5) (PCMs 8, 9), counters

Treasure hunt

Coordinates
GROUP ACTIVITY 5

3–4 children

Coordinate grid, three sets of number cards (⁻5 to 5), counters

Shuffle the cards and place them in two piles, face down.

Each write a secret list of eight points on the grid where there is some buried treasure.

Take turns to pick a card from each pile. The first pile gives you the horizontal coordinate, the second pile gives you the vertical coordinate. Place a counter on that point.

Play six rounds altogether. Compare the position of your counters with your secret lists. Has any of the treasure been found? Score one point for each treasure which has not been found.

Play again.

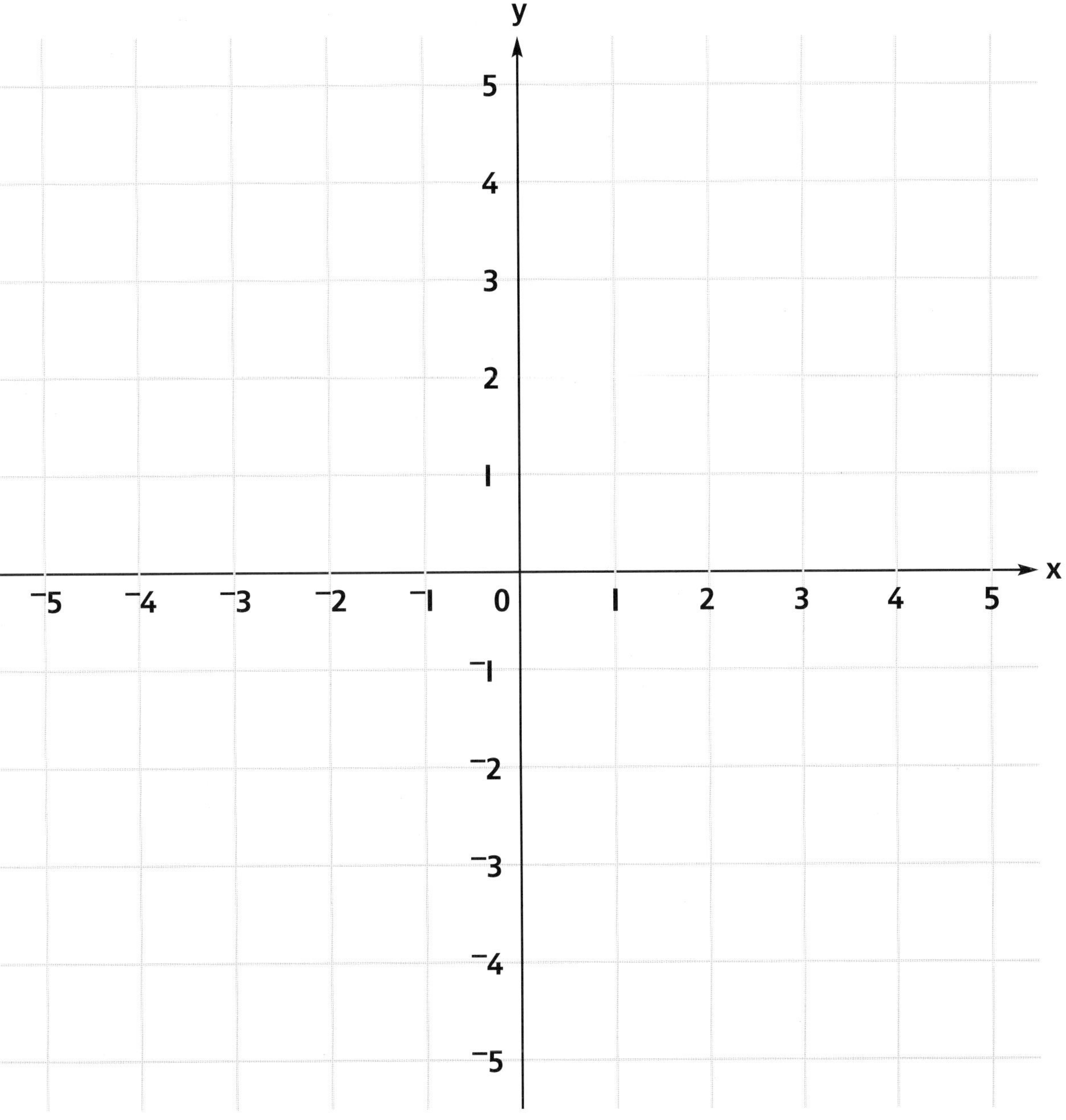

Abacus Ginn and Company 2001 Copying permitted for purchasing school only. This material is not copyright free.

S4 Reflection

ACTIVITY 1
Whole class, in pairs

- *Drawing a reflection of a shape in two mirror lines*

Squared paper, coloured pens or pencils

Each child draws a pair of axes, horizontal and vertical on some squared paper. They draw a right-angled triangle in the first quadrant of their grid, then pass the paper to their partner. They then draw a reflection of the shape in the vertical axis, and also a reflection in the horizontal axis. They pass the papers back for checking. Repeat, with different shapes and by drawing the original shape in different quadrants.

ACTIVITY 2
Pairs

- *Drawing a reflection of a shape in a mirror line*

Squared paper, rulers, a mirror

Each child draws a horizontal or vertical mirror line on some squared paper. Then they draw a rectangle of any size on one side of the mirror line, using the lines of the paper. They swap papers and each child draws a reflection of the rectangle in the mirror line. The children check each other's drawings. They use the mirror to confirm the reflections. Repeat for a right-angled triangle. Extend the activity by asking the children to draw rectangles at an angle.

ACITIVITY 3
Pairs

- *Deducing the coordinates of a shape, given the coordinates of its reflection in a named axis*

Coordinate grid (PCM 21)

Each child secretly draws a triangle on their grid, using the coordinates at its vertices, and writes the coordinates down. They then draw its reflection in the horizontal axis. They write the coordinates of the reflection on a piece of paper, and pass it to their partner to work out the coordinates of the original shape. The children check each other's answers. Repeat for different shapes drawn in different quadrants.

ACTIVITY 4
3–4 children

- *Plotting and reading points in four quadrants on a coordinate grid, investigating rules for the coordinates when points are reflected in each axis*

Two sets of number cards ($^{-}5$ to 5) (PCMs 8, 9), coordinate grid (PCM 21), interlocking cubes

Each set of cards is shuffled and placed in two piles, face down, one pile for the horizontal coordinate, one for the vertical coordinate. The children take turns to turn over one card from each pile to create a coordinate, e.g. ($^{-}3$, 2). On their turn, the children say the coordinates of the point when reflected in an axis chosen by the other children (x-axis or y-axis). They check the answer by using the coordinate grid. If they are correct, they collect a cube. The children have several turns each. Can they investigate rules for the coordinates when the points are reflected in each axis?

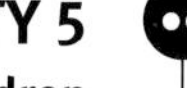

GROUP ACTIVITY 5
4 children

- *Drawing a reflection of a shape in an axis on a coordinate grid*

Coordinate grid (PCM 21)

Symmetrical shapes

4 children

Reflection
GROUP ACTIVITY 5

Coordinate grid

One of you draw a shape with straight sides in the top right of the grid.

Write the coordinates of its vertices (corners).

Another child then draws a reflection of your shape in the y-axis and writes its coordinates.

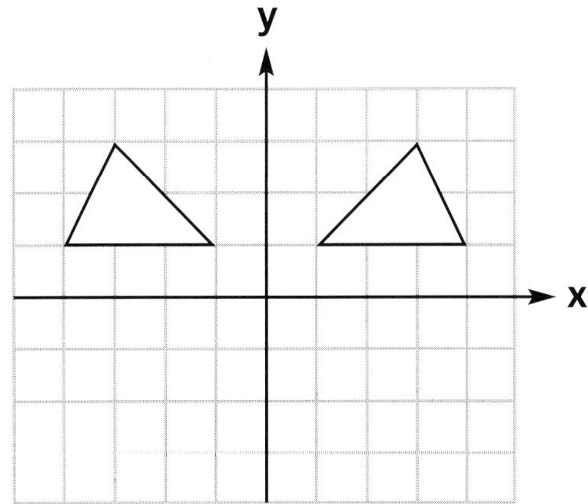

Another child then draws a reflection of the new shape in the x-axis and writes its coordinates.

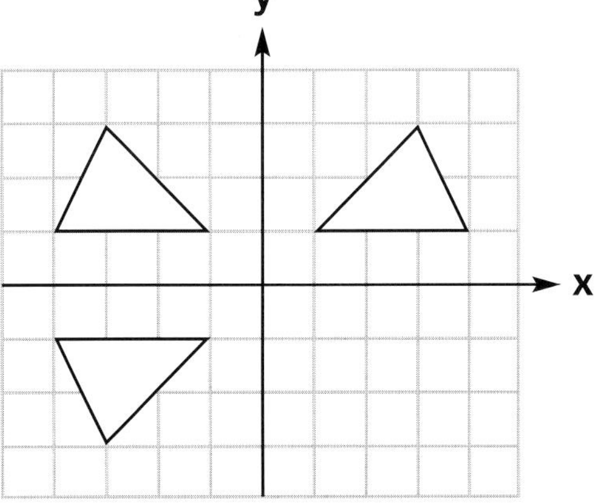

The fourth child then draws a reflection of the last shape in the y-axis and writes its coordinates.

Repeat, taking turns to start.

Abacus Ginn and Company 2001 Copying permitted for purchasing school only. This material is not copyright free.

S5 Rotation and translation

ACTIVITY 1
Whole class, in pairs

- *Drawing a rotation, about a given point, of a shape which has a vertex on the point*

Squared paper, pencils

Each child draws a rotation point on some squared paper. They then draw a rectangle which has this point at one of its vertices. They pass it to their partner. They then draw rotations of the shape about the points of 90°, 180° and 270°. They pass the papers back to be checked. Repeat with different shapes, each with one vertex on the point of rotation.

ACTIVITY 2
2–3 children

- *Describing different translations of shapes on a square grid*

Card, cm-squared paper, scissors

The children cut out a 4 cm × 6 cm card rectangle and draw round it on the squared paper. They then slide it to different positions on the paper, and write inside each rectangle the translation from the original position, e.g. '5 cm right, 2 cm up'. Repeat, using a different starting shape such as a right-angled triangle.

ACTIVITY 3
Pairs

- *Deducing the coordinates of a shape, given its coordinates after a defined rotation*

Coordinate grid (PCM 21)

Each child secretly draws a rectangle on their grid, using the coordinates at its vertices, and writes the coordinates down. They mark one of its vertices as a centre of rotation. They then draw its position after a rotation of 90° clockwise. They write the coordinates of the rotated shape on a piece of paper, and pass it to their partner, who has to work out the coordinates of the original shape. The children check each other's answers. Repeat with different shapes drawn in different quadrants, and with different rotations.

ACTIVITY 4
3–4 children

- *Plotting and reading points in four quadrants on a coordinate grid, investigating rules for the coordinates when points are rotated about the origin*

Two sets of number cards (⁻5 to 5) (PCMs 8, 9), coordinate grid (PCM 21), interlocking cubes

Each set of cards is shuffled and placed in two piles, face down, one pile for the horizontal coordinate, one for the vertical coordinate. The children take turns to turn over one card from each pile to create a coordinate, e.g. (4, ⁻1). On their turn, the children have to say the coordinates of the point after a rotation chosen by the others, e.g. 90° anticlockwise. They check the answer by using the coordinate grid. If they are correct, they collect a cube. Repeat several times. Let them investigate rules for the coordinates when points are given rotations of 90°, 180°, 270° clockwise.

GROUP ACTIVITY 5
3–4 children

- *Recognising and describing rotations*

Scissors

Rotation cards

3–4 children

Scissors

Cut out these rotation cards. Spread them out face down on the table.

Take turns to turn over a card and describe the rotation from the shaded shape to the clear shape about the marked point, for example, a rotation of 90° clockwise.

Continue until all the cards have been turned over.

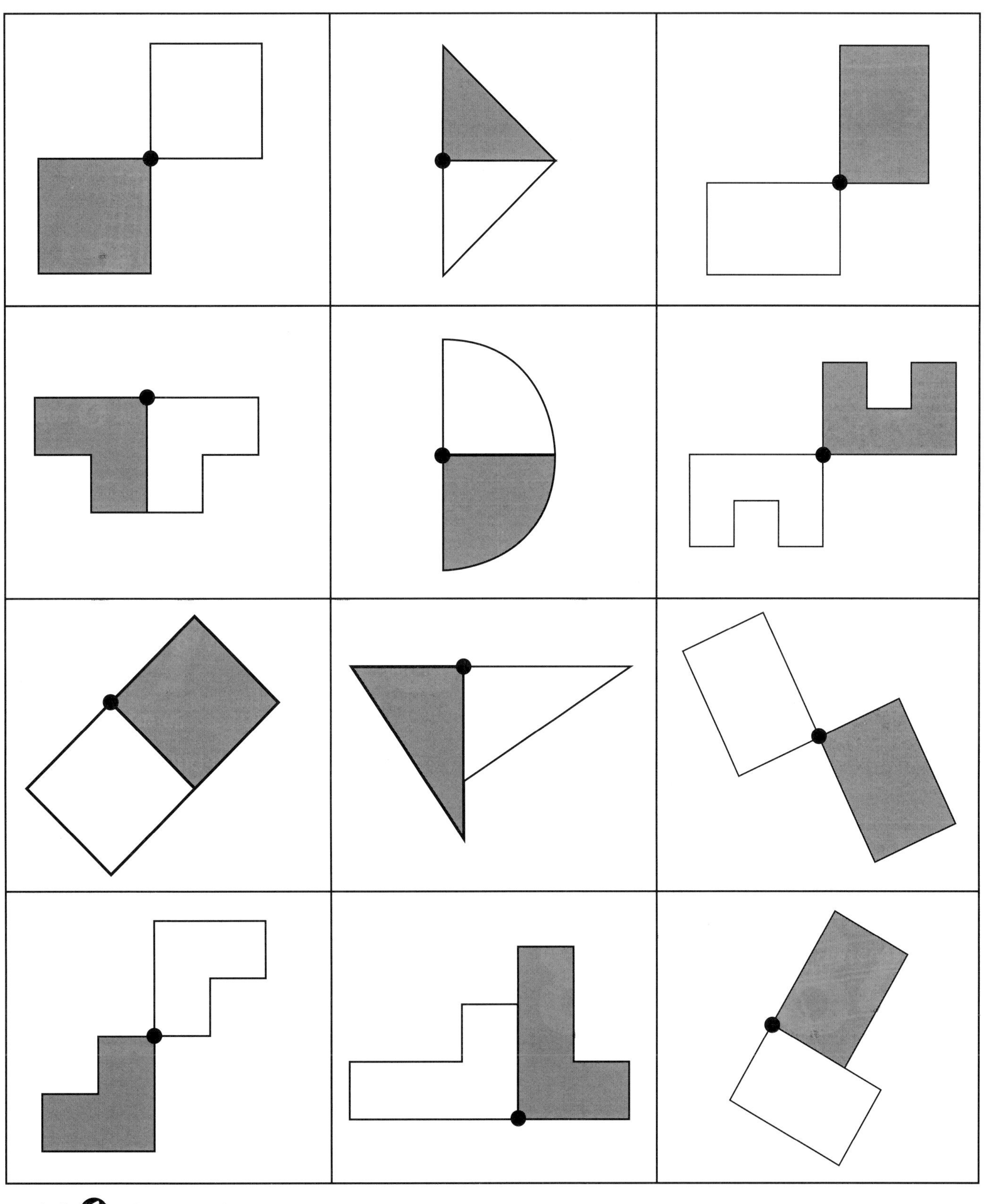

Abacus Ginn and Company 2001 Copying permitted for purchasing school only. This material is not copyright free.

S6 3-d shape

ACTIVITY 1
Whole class, in pairs

• *Recognising parallel faces and the names and properties of 3-d shapes*

A set of 3-d shapes (with flat faces), a cloth bag, cards (labelled 'yes' on one side, 'no' on the other)

Put the shapes in the bag, and ask a child to take one out. Hold it up to show the class. The children decide if it has parallel faces or not, by holding up 'yes' or 'no' on their cards. Discuss the reasons together. Each pair then writes down the name of the shape. Check that they are correct. Repeat for each shape in the bag.

ACTIVITY 2
3 children

• *Recognising parallel and perpendicular faces in a 3-d shape*

A set of 3-d shapes (with flat faces), a cloth bag

Place the shapes in the bag. One child takes a shape from the bag and points to one of its faces. The second child points out a parallel face (if possible). The third child points out a perpendicular face (if possible). Continue until all the shapes have been removed from the bag, sharing the roles. Put the shapes back and play again.

ACTIVITY 3
3–4 children

• *Recognising the number of parallel faces in a 3-d shape*

A set of 3-d shapes (with flat faces), a cloth bag, counters

Put the shapes in the bag. In turn, each child takes a shape from the bag, and says how many pairs of parallel faces it has. They check each other's answers. If they are correct, they take a matching number of counters. Continue until all the shapes have been removed from the bag. Repeat. Who has the most counters?

ACTIVITY 4
3 children

• *Recognising the names and properties of 3-d shapes*

A set of 3-d shapes (with flat faces), a cloth bag, counters

One child holds up one of the shapes. Each child writes down its name. The children receive one counter if they are correct. They discuss its properties together, including the properties of its faces, e.g. does it have parallel faces, perpendicular faces? Repeat for each shape in turn.

GROUP ACTIVITY 5
2–3 children

• *Recognising the properties of the faces and the names of 3-d shapes*

A set of 3-d shapes (with flat faces), a cloth bag

GROUP ACTIVITY 6
3–4 children

• *Recognising the number of parallel faces in a 3-d shape*

Several sets of number cards (0 to 10) (PCM 9), Polydron or Clixi

Recognising faces

3-d shape
GROUP ACTIVITY 5

2–3 children

A set of 3-d shapes, a cloth bag

Place the shapes in the bag.

Take turns to pick a shape. The player on your left has to say:

1. the number of faces
2. the number of pairs of parallel faces (if any)
3. the name of the shape.

Check each other's descriptions.

When all the shapes have been described, play again.

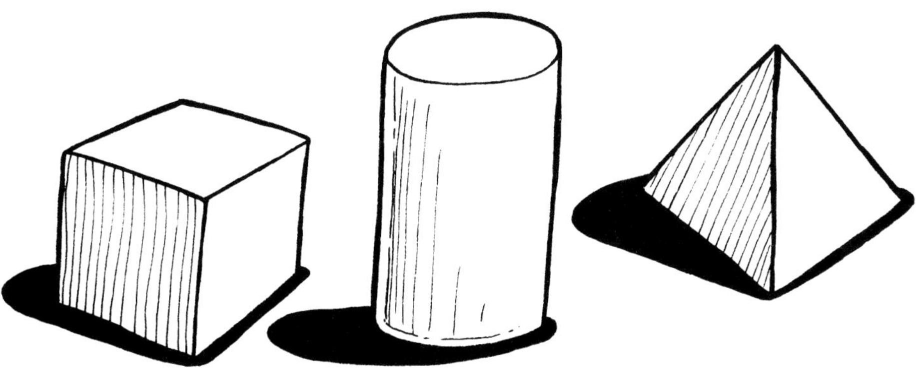

Parallel faces

3-d shape
GROUP ACTIVITY 6

3–4 children

Several sets of number cards (0 to 10), Polydron or Clixi

Build lots of different shapes with parallel faces.

Place a number card beside each one to show how many pairs of parallel faces it has.

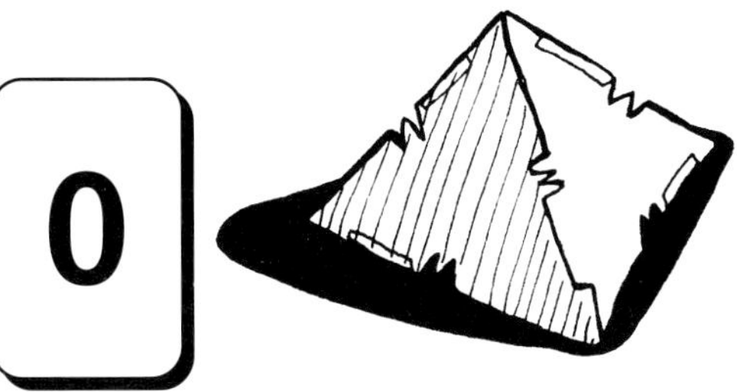

Abacus Ginn and Company 2001 Copying permitted for purchasing school only. This material is not copyright free.

S7 2-d shape

ACTIVITY 1
Whole class, in pairs

- *Constructing 2-d shapes on a geoboard*

A geoboard, rubber bands, shape labels ('square', 'triangle', 'rectangle', 'parallelogram', 'rhombus', 'trapezium', 'pentagon') (PCM 28), counters

Choose a label at random, e.g. parallelogram. Each pair constructs a parallelogram on their geoboard. They hold them up for others to see and check. If they are correct, they collect a counter. Continue for each label in turn. If it is impossible to make the shape, for example, a rhombus (the distance between diagonal pegs is longer than that between horizontal and vertical pegs), the children must say so; such statements are also worth a counter. Which pair has collected the most counters?

ACTIVITY 2
2+ children

- *Constructing shapes with parallel sides*

Squared paper

The children draw six different parallelograms on the squared paper. For each shape, they draw the parallel sides in the same colour. The children check each other's drawings.

ACTIVITY 3
3–4 children

- *Recognising shapes with parallel sides, naming 2-d shapes*

Pinboard shape cards (PCMs 22, 23), counters

Shuffle the cards and spread them out face down. The children take turns to reveal a card and say how many pairs of parallel sides it has. They then say the name of the shape. They check each other's answers. If they are correct, they take a matching number of counters, e.g. if it has one pair of parallel sides, they take one counter. Continue until all the cards have been turned over. Who has the most counters? Repeat several times.

ACTIVITY 4
2–3 children

- *Constructing and naming 2-d shapes*

3 × 3 geoboard, rubber bands, geoboard shape paper (PCM 24)

The children use the rubber bands to create different shapes on the geoboard. They then copy each shape onto the geoboard shape paper and write the name of the shape underneath. They check each other's answers.

GROUP ACTIVITY 5
3–4 children

- *Recognising the names and properties of 2-d shapes*

Shape labels ('square', 'triangle', 'rectangle', 'parallelogram', 'rhombus', 'trapezium', 'pentagon') (PCM 28), pinboard shape cards (PCMs 22, 23), counters

GROUP ACTIVITY 6
2–3 children

- *Constructing and naming shapes with parallel sides*

3 × 3 geoboard, rubber bands, geoboard shape paper (PCM 24)

Parallel sides and right angles

2-d shape
GROUP ACTIVITY 5

3–4 children

Shape labels, pinboard shape cards, counters

Shuffle the cards and place them face down in a pile.

Take turns to take a card and place it beside its matching label.

Score one point for each pair of parallel sides on the shape, and one point for each right angle.

Check each other's scoring.

Collect counters to match the points scored.

When all the cards are taken, who has the most points?

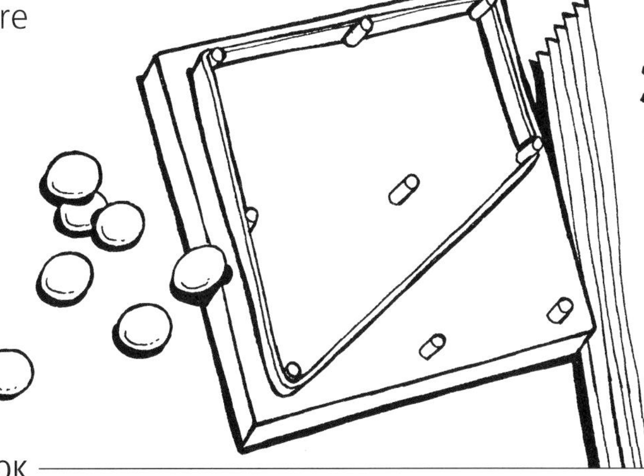

Creating shapes with parallel sides

2-d shape
GROUP ACTIVITY 6

2–3 children

3×3 geoboard, rubber bands, geoboard shape paper

Make different shapes on the geoboard, each with parallel sides.

Draw them on the geoboard paper.

How many different shapes can you make?

Name each one.

Now make shapes on a 4×4 geoboard.

Draw them on geoboard paper.

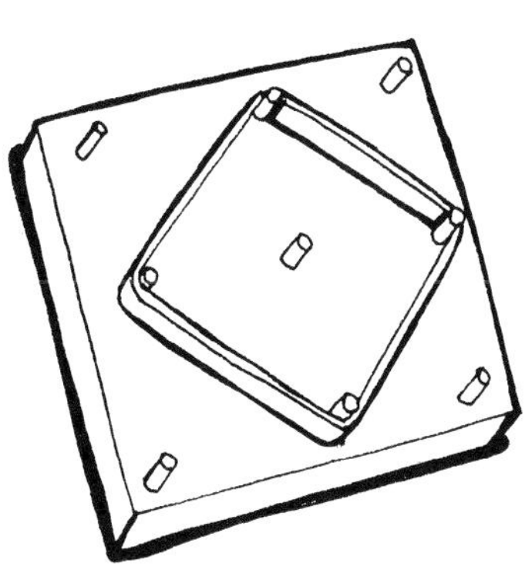

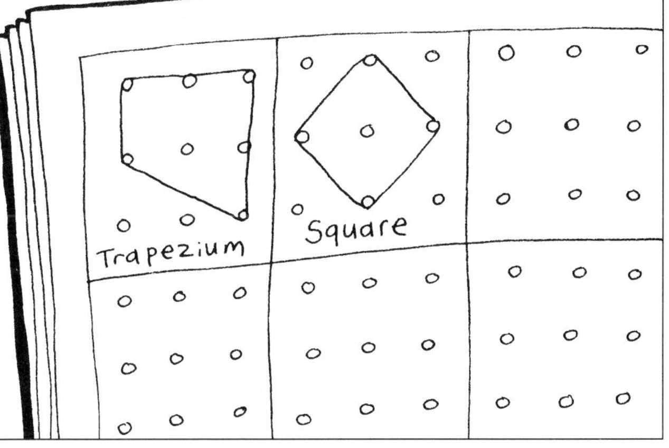

Abacus Ginn and Company 2001 Copying permitted for purchasing school only. This material is not copyright free.

2-d shape

ACTIVITY 1
Whole class, in pairs

- *Recognising and naming different quadrilaterals*

A set of different 2-d quadrilaterals (square, rectangle, parallelogram, trapezium, rhombus, kite), counters

Write the names of the shapes on the board. Hold up a shape at random. The children decide on its name, and write it down. Discuss the correct answer. Pairs with a correct answer collect a counter. Discuss some of the properties of the shape. Repeat for each shape in turn. Which pairs collected a maximum number of counters?

ACTIVITY 2
2+ children

- *Recognising and naming different quadrilaterals*

Pinboard shape cards (PCMs 22, 23)

The children sort the shapes according to their number of sides, i.e. into triangles, quadrilaterals, pentagons, etc. They then sort the quadrilaterals into different types, naming each set. They describe the properties of the shapes in each set.

ACTIVITY 3
Pairs

- *Describing and constructing different quadrilaterals*

2-d quadrilaterals, a cloth bag

One child secretly takes a shape from the bag. They describe the shape, without naming it, so that their partner can draw it. The children compare the finished drawing with the shape. Repeat several times, with the children sharing the roles.

ACTIVITY 4
2–3 children

- *Investigating the properties of the diagonals of types of quadrilateral*

2-d quadrilaterals, rulers

The children draw around the different quadrilaterals, then draw the diagonals. They investigate the properties of the diagonals of different quadrilaterals, e.g. do they bisect each other? Do they meet at right angles? Are the diagonals equal?

ACTIVITY 5
3–4 children

- *Investigating the properties of the angles of types of quadrilateral*

Quadrilateral cards (see Group Activity 6)

The children describe the properties of the angles of each type of quadrilateral. For example, how many of them have right angles? How many right angles do they have? How many have all acute angles? How many have obtuse angles? Do any have a reflex angle?

GROUP ACTIVITY 6
3–4 children

- *Recognising the names and properties of different quadrilaterals*

Scissors

Quadrilateral cards

2-d shape
GROUP ACTIVITY 6

3–4 children

Scissors

Cut out these quadrilateral cards. Spread them out face down on the table.

Take turns to turn over a card and name the shape.

Describe its particular properties, referring to its sides and angles.

1	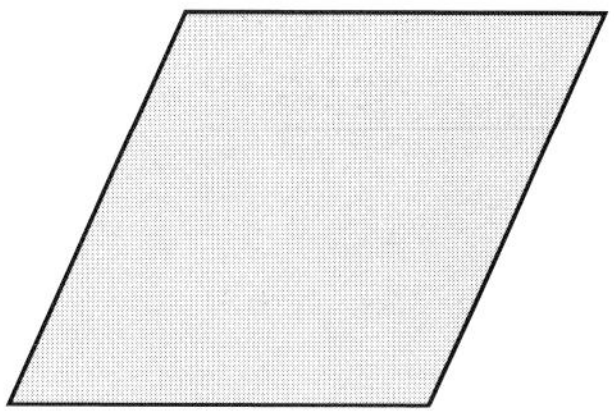 2	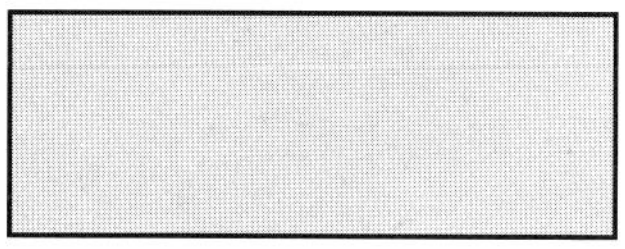 3
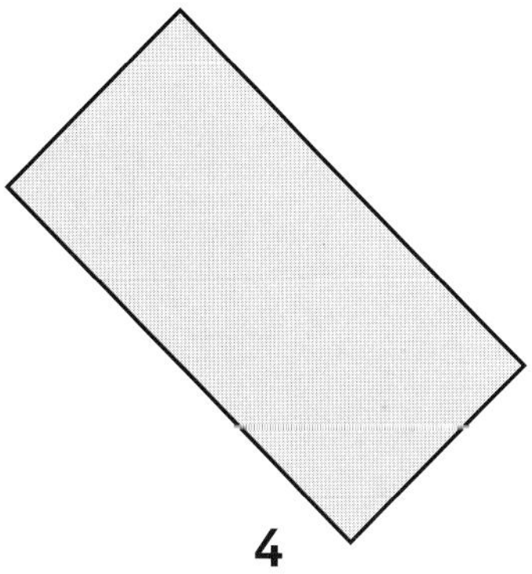 4	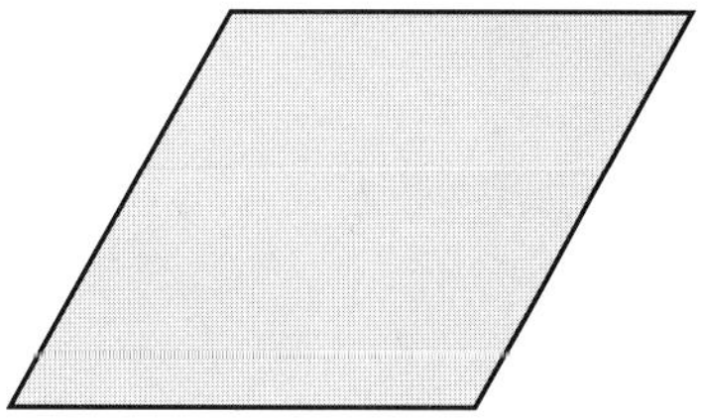 5	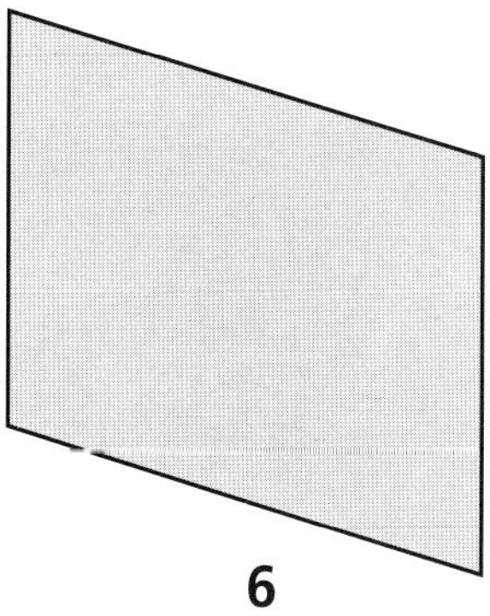 6
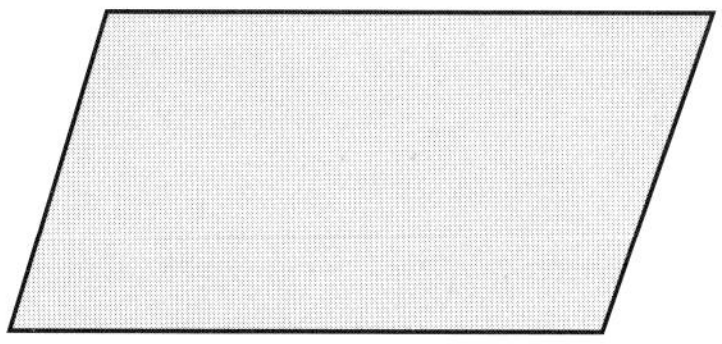 7	8	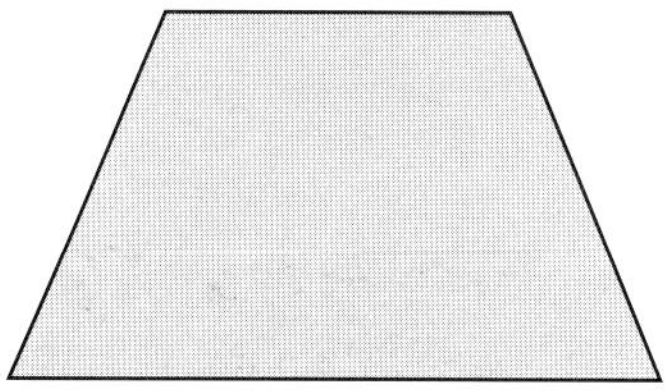 9
10	11	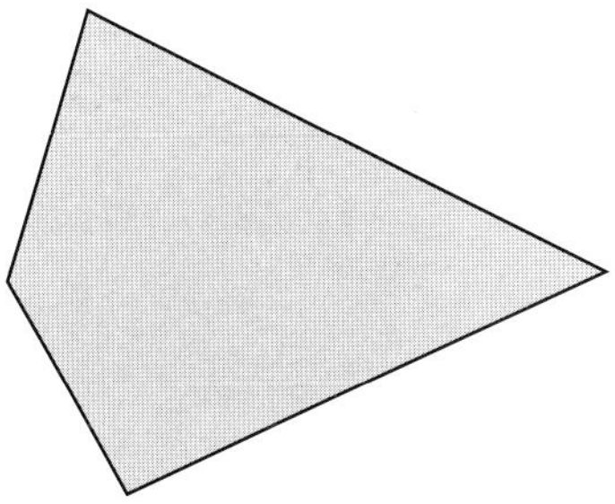 12

Abacus Ginn and Company 2001 Copying permitted for purchasing school only. This material is not copyright free.

S9 2-d shape

ACTIVITY 1
Whole class, in pairs

• *Recognising and naming different polygons, recognising regular and irregular polygons*

A set of polygons, a cloth bag, cards (labelled 'yes' on one side, 'no' on the other)

Put the shapes in the bag, and choose a child to take one out. Hold it up for the class to see. The children decide if it is a regular polygon or not, by showing 'yes' or 'no' on their cards. Discuss the reasons together. Each pair then writes down the name of the polygon. Check that they have the correct name. Repeat until the bag is empty.

ACTIVITY 2
2+ children

• *Recognising and naming different quadrilaterals and polygons*

Pinboard shape cards (PCMs 22, 23), shape labels ('triangle', 'square', 'rectangle', 'parallelogram', 'trapezium', 'kite', 'pentagon', 'hexagon') (PCMs 28, 29)

The children spread out the labels, face up. They shuffle the cards and place them face down in a pile. They take turns to pick a card and place it beside its matching shape label. They check each other's decisions. How many shape cards are in each set? Reshuffle the cards and play again.

ACTIVITY 3
2–3 children

• *Recognising the names of different polygons*

2-d polygons, shape labels ('triangle', 'quadrilateral', 'pentagon', 'hexagon', 'heptagon', 'octagon', 'nonagon', 'decagon') (PCMs 28, 29), a cloth bag

Place the shapes in the bag. One child, in turn, takes a polygon from the bag, decides on its name, and places it alongside its matching label. The children check each other's decisions. Repeat for each shape.

ACTIVITY 4
2–3 children

• *Recognising the properties of polygons*

Pinboard shape cards (PCMs 22, 23)

The cards are shuffled and placed face down in a pile. The children take turns to pick a card. They say the name of the shape and then say five facts about it, such as the number of sides, vertices, angles, right angles, and parallel sides. The children check each other's statements. When all the cards have been taken, they reshuffle them and play again.

GROUP ACTIVITY 5
3–4 children

• *Constructing polygons using the pieces of a dissection of a rectangle*

A rectangle (photocopied onto card), scissors

GROUP ACTIVITY 6
3–4 children

• *Constructing polygons using the pieces of a dissection of a square*

A square (photocopied onto card), scissors

Rectangle pieces

2-d shape
GROUP ACTIVITY 5

3–4 children

A rectangle (photocopied onto card), scissors

Cut out the rectangle, then cut along the lines to make three pieces. Make shapes by joining the pieces along sides of equal length.

Can you join two of the pieces to make these shapes: a triangle, a trapezium? Draw and label each of them.

Can you join three of the pieces to make these shapes: a parallelogram, a trapezium, a kite, a pentagon? Draw and label each of them.

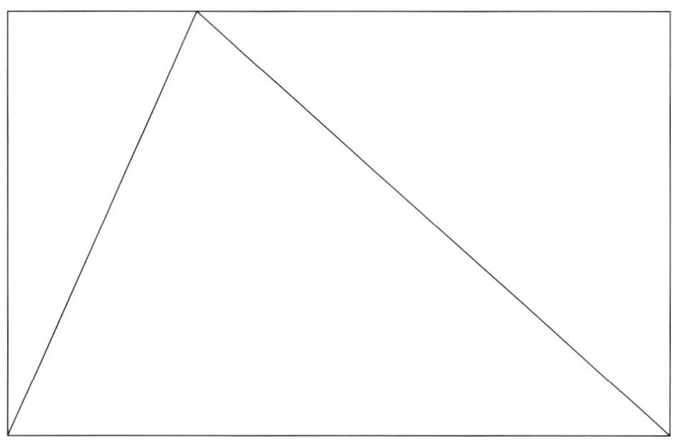

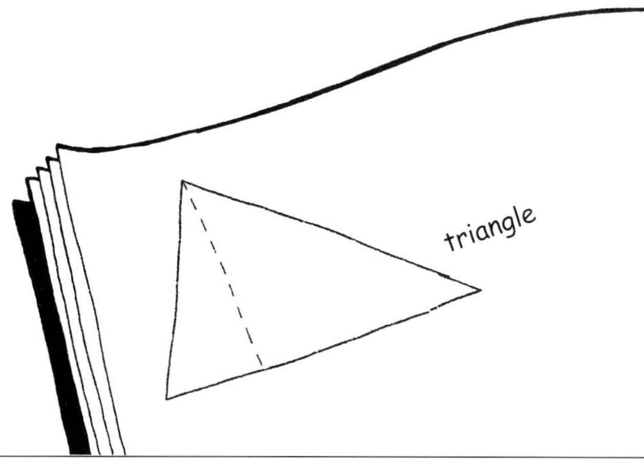

Square pieces

2-d shape
GROUP ACTIVITY 6

3–4 children

A square (photocopied onto card), scissors

Cut out this square, then cut along the lines to make three pieces.

Make shapes by joining the pieces along sides of equal length.

Investigate how many different shapes you can make.

Draw and label each of them.

triangle

Abacus Ginn and Company 2001 Copying permitted for purchasing school only. This material is not copyright free.

D1 Grouped data

ACTIVITY 1
Whole class, in pairs

- *Constructing and interpreting a grouped frequency table, drawing a bar graph to represent a grouped frequency table*

Lists of class first names, graph paper

The children write the number of letters in the first names of everyone in the class. They complete a grouped frequency table based on the number of letters which should be grouped 1–3, 4–6, 7–9 and 10+ letters. They draw a bar graph to illustrate the results and then write about the results.

ACTIVITY 2
2 pairs

- *Constructing and interpreting a grouped frequency table, drawing a bar graph to represent a grouped frequency table*

Two dice (1 to 6)

The children draw a grouped frequency table based on the total of two dice being rolled. The ranges of total scores should be grouped 1–3, 4–6, 7–9 and 10–12. Each pair chooses two of the ranges. The two dice are rolled, the scores added and a tally mark placed against the correct range in the table. The children continue to roll the dice until one range has a frequency of 20 rolls. The winning pair is the one who chose the highest scoring range. Finally, the children add up all the totals and draw a bar graph to represent the results.

GROUP ACTIVITY 3
3–4 children

- *Constructing and interpreting a grouped frequency table*

Two sets of number cards (1 to 10) (PCM 9), a calculator

GROUP ACTIVITY 4
Pairs

- *Constructing and interpreting a grouped frequency table*

National football results

D1

Tally times

Grouped data
GROUP ACTIVITY 3

3–4 children

Two sets of number cards (1 to 10), a calculator

Shuffle the cards separately and place them face down in two piles.

Draw a grouped frequency table like this:

Answer to multiplication	Tallies	Frequency
0–20		
21–40		
41–60		
61–80		
81–100		

Take turns to pick a card from each pile and multiply the two numbers together. Put a tally mark in the table against the range your answer falls in.

Continue, reshuffling the cards after ten rounds.

If you need to, use a calculator to check your multiplications.

Stop when one group has a frequency of 15.

Before you start, all guess which group will win.

D1

Goal times

Grouped data
GROUP ACTIVITY 4

Pairs

National football results

Look at the times in each match when goals are scored.

Draw and complete this grouped frequency table.

Minute of goal	Tallies	Frequency
1–15		
16–30		
31–45		
46–60		
61–75		
76–90		

Write about the results.

Abacus Ginn and Company 2001 Copying permitted for purchasing school only. This material is not copyright free.

D2 Pie charts

ACTIVITY 1

Whole class, in pairs

- *Constructing and interpreting a pie chart*

8-point pie chart (PCM 25), interlocking cubes (four colours), a cloth bag
The children place all the cubes in the bag, shake the bag and take out eight cubes. They then draw a pie chart to show how many cubes of each colour have been taken out. The children colour the slices of the pies to match the colours of the cubes. Discuss the results. Which colour was most common? Least common? How many more red cubes than blue were there? Repeat, removing 16 cubes instead of eight.

ACTIVITY 2

2–3 children

- *Constructing and interpreting a pie chart*

8-point pie chart (PCM 25), 16-point pie chart (PCM 27)
The children collect data from eight other people about their favourite colour. They draw a pie chart to show the results. They then collect data from another eight people and draw a new pie chart based on all 16 votes. They then compare the two charts.

ACTIVITY 3

3–4 children

- *Constructing and interpreting a pie chart*
- *Using computer software to process data*

Spreadsheet software or graphing package
The children conduct a simple survey based on favourite television programmes. They enter the data into a computer spreadsheet. They then use the data to construct a pie chart. They write a short report about the pie chart on a word processor. Repeat for favourite ice-cream flavour, favourite colour, etc.

GROUP ACTIVITY 4

3–4 children

- *Constructing and interpreting a pie chart*

A pack of playing cards, 16-point pie chart (PCM 27)

GROUP ACTIVITY 5

3–4 children

- *Constructing and interpreting a pie chart*

12-point pie chart (PCM 26)

Playing cards

Pie charts
GROUP ACTIVITY 4

3–4 children

A pack of playing cards, 16-point pie chart

Shuffle the cards and deal out 16.

Draw a pie chart to show how many of each suit (hearts ♥, clubs ♣, diamonds ♦, or spades ♠) you have dealt.

Repeat the activity with 32 cards.

How do you spend your day?

Pie charts
GROUP ACTIVITY 5

3–4 children

Each write down how many hours, out of 24 in a day, you spend:

(a) sleeping
(b) playing
(c) at school
(d) watching TV
(e) eating
(f) doing something else.

Draw a pie chart to show the number of hours you spend doing each thing.

Compare each other's charts.

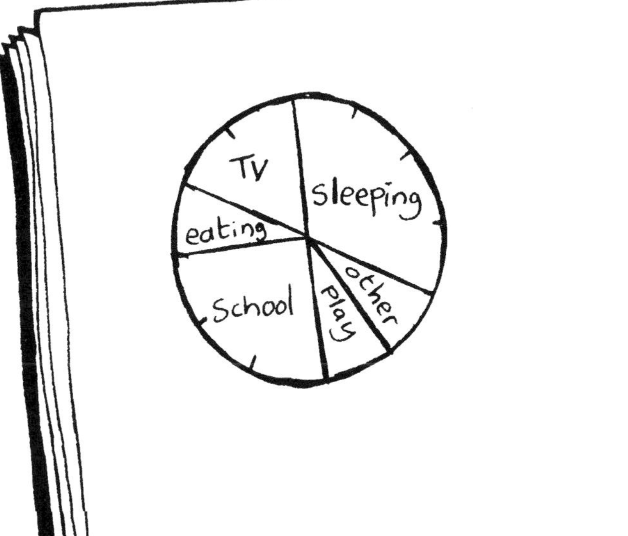

Abacus Ginn and Company 2001 Copying permitted for purchasing school only. This material is not copyright free.

D3 Conversion graphs

ACTIVITY 1
Whole class, in pairs

• *Constructing and interpreting a conversion graph*

Graph paper (PCM 30), ten small objects (less than 30 cm long)

The children draw a graph to convert inches into centimetres. They label the horizontal axis 'inches', and mark it in units of 1 inch up to 12 inches. They label the vertical axis 'centimetres', and mark it in units of 5 centimetres up to 30 centimetres. They mark the point on the graph to show 30 cm = 12 inches. They draw the conversion line from (0, 0) to the point. They then measure the length in centimetres of ten objects and use the graph to convert each measurement into inches.

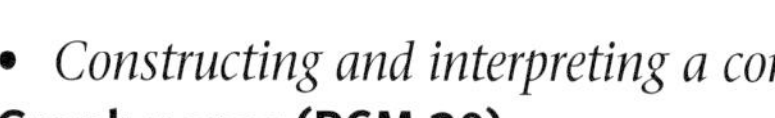

ACTIVITY 2
2–3 children

• *Constructing and interpreting a conversion graph*

Graph paper (PCM 30)

The children draw a graph to convert pounds into another currency. They label the horizontal axis 'pounds' and mark it in units of £10 up to £100. They label the vertical axis according to the chosen currency and mark it in appropriate units. They calculate the value of £100 in their chosen currency. They plot this point on the graph, and draw a conversion line from (0, 0) to the point. They use this conversion line to convert from one unit to the other. They find the value of £10, £25, £50, £75 etc in the other currency.

GROUP ACTIVITY 3
3–4 children

• *Using a conversion graph to convert between grams and ounces*

Weighing scales

Grams and ounces

3–4 children

Weighing scales

Find eight different objects that weigh less than 500 g (pencil cases, building blocks, scissors, books etc).

Weigh each object in grams, then use the graph to convert each weight into ounces.

Record the weights in a table.

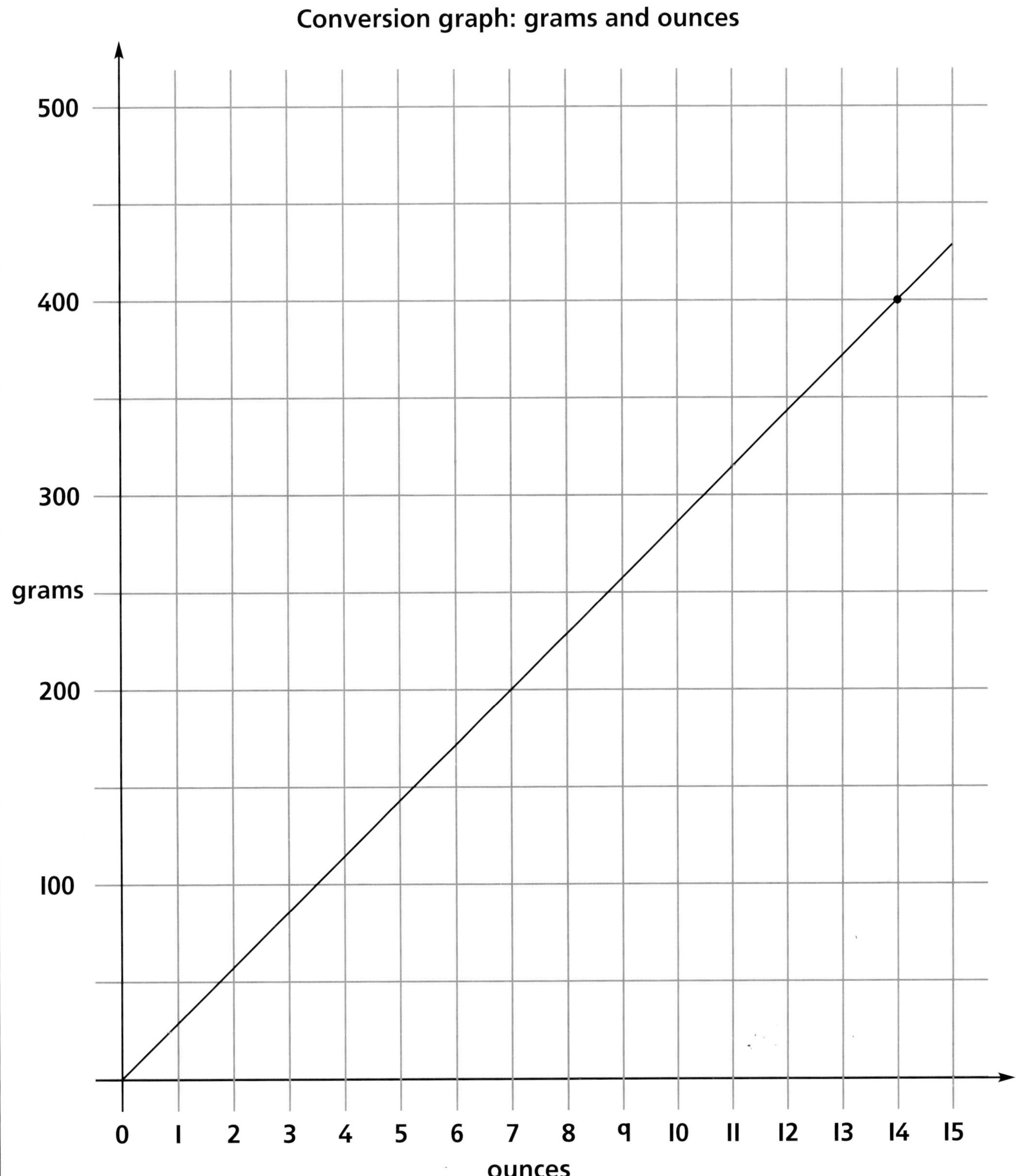

Abacus Ginn and Company 2001 Copying permitted for purchasing school only. This material is not copyright free.

D4 Averages

ACTIVITY 1
Whole class, in pairs

- *Calculating the mean of a set of three or four numbers up to 10*

Several sets of number cards (1 to 10) (PCM 9), interlocking cubes
Shuffle the cards and deal three to each child. They find the average of the three cards, and collect cubes to match the whole number part of the average. Repeat for ten rounds. Who has the most cubes? Repeat, dealing four cards each.

ACTIVITY 2
2–4 children

- *Calculating the mean of an amount of money*

Coins (1p and 10p)
Starting with 1p coins only, two children take a handful of coins each. They find the mean amount taken. Repeat for three children, then for four. Repeat each activity, this time using both 10p and 1p coins.

ACTIVITY 3
2–4 children

- *Calculating a mean and mode*

Two reading books
The children open a book at the beginning of a chapter, and record the number of letters in each word for the first twenty words. They find the mean and mode for the number of letters per word. They repeat for a different book and compare the two results.

ACTIVITY 4
3–4 children

- *Calculating a mean, median and mode*

Game 8: 'Seeing Stars!', counters, a dice (1 to 6)

GROUP ACTIVITY 5
3–4 children

- *Calculating a mean and median*

A list of national football results

GROUP ACTIVITY 6
3–4 children

- *Calculating a mean and median*

Two dice (1 to 6)

Goals

Averages
GROUP ACTIVITY 5

3–4 children

National football results

Choose ten football teams and write down how many goals they each scored.

Find the mean and median number of goals scored by the teams.

Repeat the same activity for 20 teams, then 30 teams.

How do the means and medians compare?

Manchester Utd - 3
Liverpool - 2
Blackburn Rovers - 1
Leicester City - 4

Dice differences

Averages
GROUP ACTIVITY 6

3–4 children

Two dice (1 to 6)

Roll two dice and write the difference between each number of spots.

Do this ten times.

Find the mean and median difference.

Repeat the activity for 20 rolls, then 40.

Do the means and medians change?

Extend to rolling the two dice and finding the total.

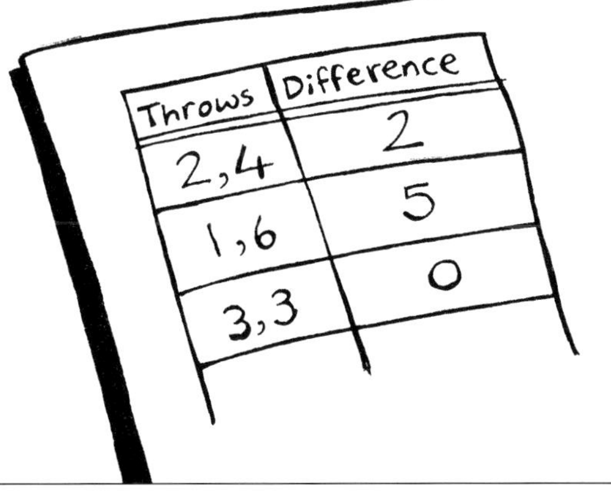

Throws	Difference
2,4	2
1,6	5
3,3	0

Abacus Ginn and Company 2001 Copying permitted for purchasing school only. This material is not copyright free.

D5 Probability

ACTIVITY 1
Whole class, in pairs

- *Predicting the outcomes of events*

Interlocking cubes (red, yellow, blue), a cloth bag

The children place four red, two blue and one yellow cube in the bag. They draw a tally chart to record the colours of cubes drawn from the bag. Before they start, they each predict how many of each colour will be taken out of the bag. They take turns to take a cube from the bag and replace it, noting and recording on the table its colour. They do this 24 times altogether. They compare the results with their predictions. The children change the proportions of colours of cubes in the bag and repeat the activity.

ACTIVITY 2
3–4 children

- *Predicting the outcomes of events*

A dice (1 to 6)

The children draw a tally chart to record the roll of a dice. The rolls are grouped as follows: 1, 2 or 3; 4 or 5; and 6. Before they start, they each predict how many rolls within each group will be thrown. The children take turns to roll the dice, recording the number in the tally chart. They do this 30 times altogether. They compare the results with their predictions. Play again, changing the groups, such as 1 or 2; 3 or 4; and 5 or 6.

GROUP ACTIVITY 3
3–4 children

- *Predicting the outcomes of events*

Event cards, a dice (1 to 6), counters

GROUP ACTIVITY 4
3–4 children

- *Predicting the outcomes of events*

Event cards, interlocking cubes (red, yellow, green) a cloth bag

Guessing game

Probability
GROUP ACTIVITY 3

3–4 children

Event cards, a dice (1 to 6), counters

Make nine 'event' cards as shown.

Shuffle them and place them face down in a pile.

Each take a card. Guess how many times a dice will match your card if you throw it 12 times.

Write down your guesses.

Throw the dice 12 times. Record the scores by placing a counter on a card each time it matches the throw.

Compare the results with your guesses.

Repeat the activity for 18 throws, or 24 throws.

⚀	⚁	⚅
odd number	even number	3 or more
more than 4	less than 3	5 or less

Colour guessing

Probability
GROUP ACTIVITY 4

3–4 children

Event cards, interlocking cubes (red, yellow, green), a cloth bag

Make nine 'event' cards as shown.

Put ten different coloured cubes in the bag and make a note of the colours.

In turn, each choose an event card.

Guess how many times your event will happen if a cube is taken out and replaced ten times.

Write down your guesses.

In turn, each take a cube out of the bag and replace it ten times.

Record each event. Compare the results with your guesses.

Repeat the activity for taking a cube 20 times.

Repeat for a different selection of cubes in the bag.

blue	green	yellow
red	not blue	not red
green or yellow	red or blue	yellow or blue

Abacus Ginn and Company 2001 Copying permitted for purchasing school only. This material is not copyright free.

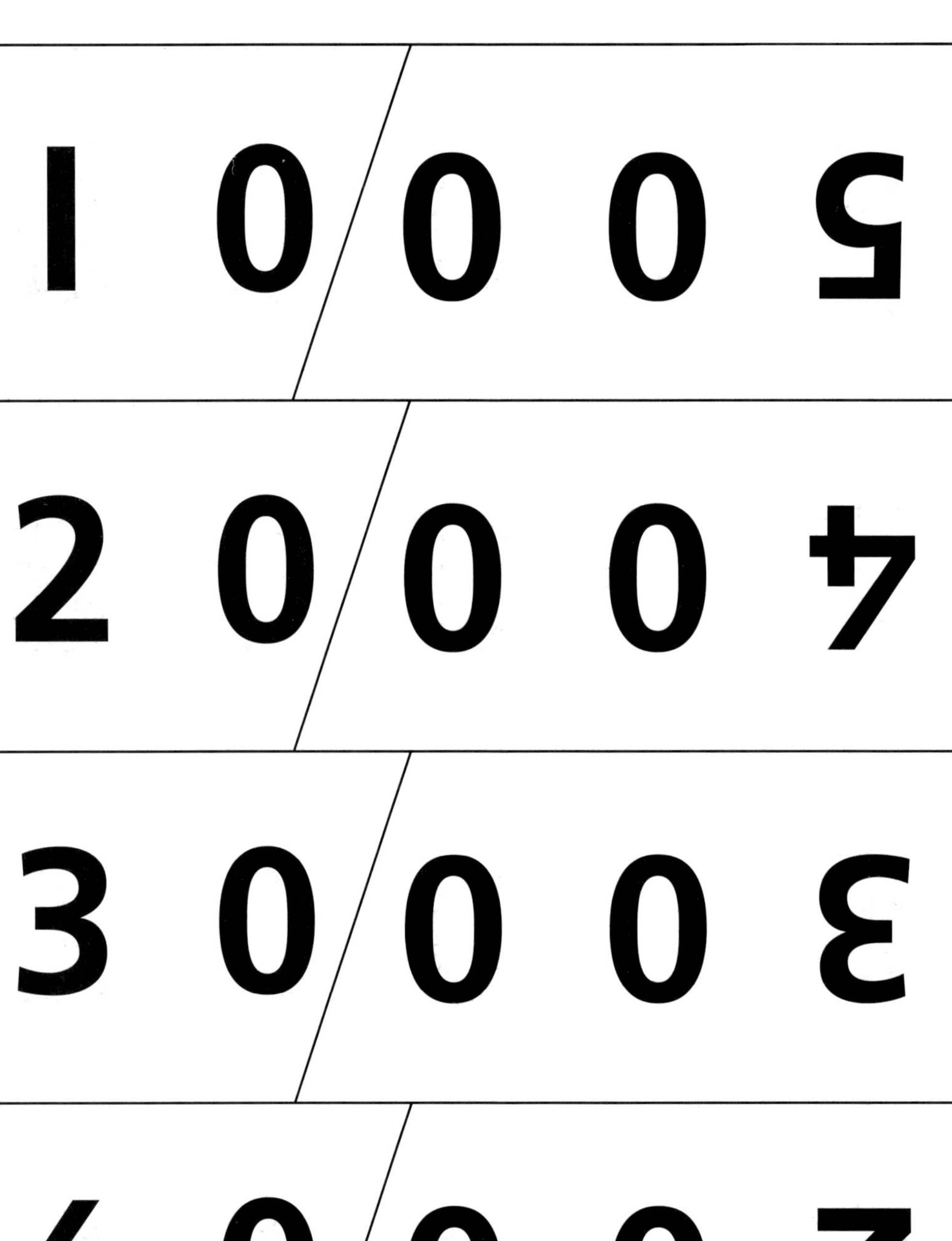

Abacus Ginn and Company 2001. Copying permitted for purchasing school only. This material is not copyright free.

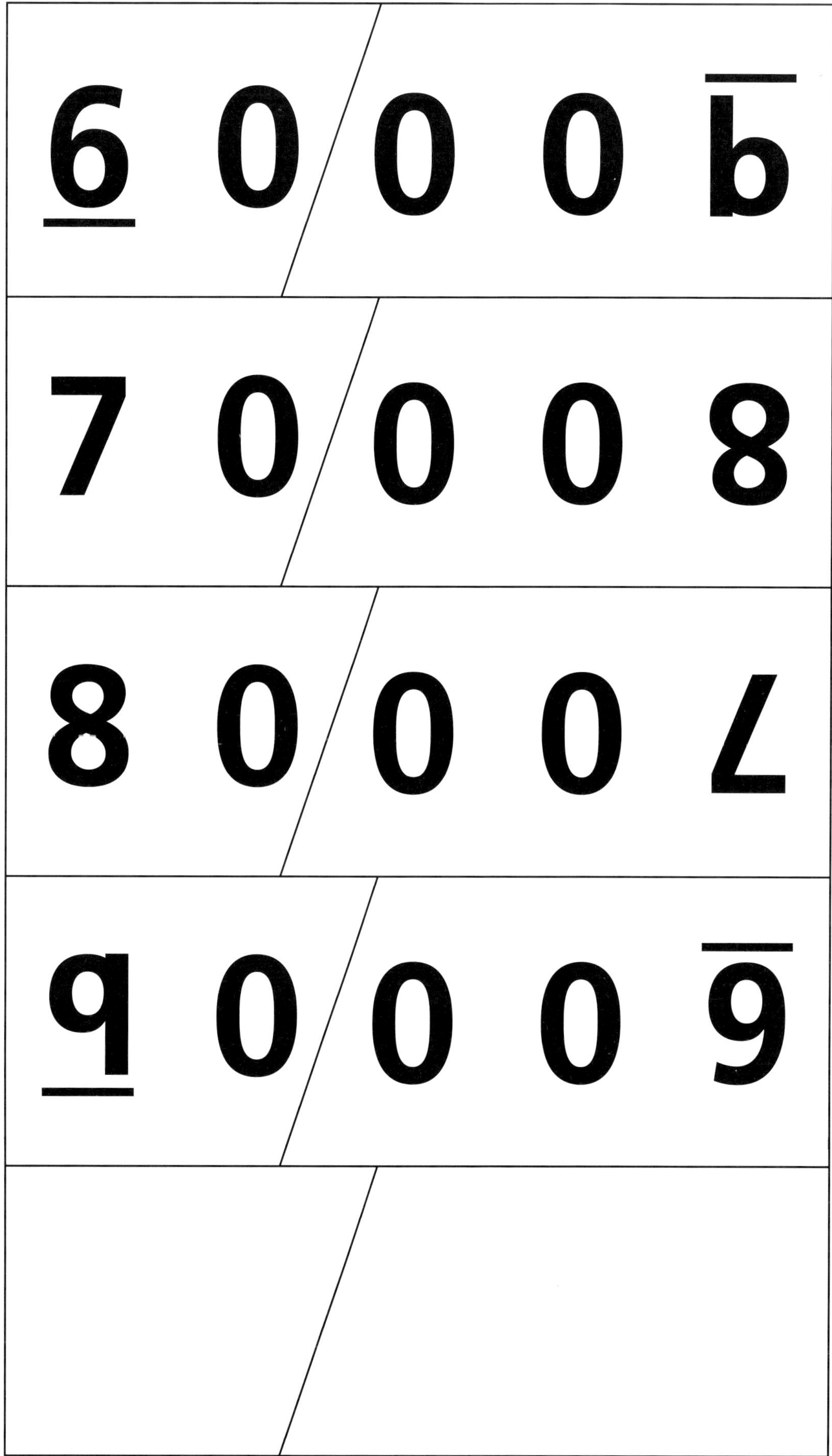

Abacus Ginn and Company 2001. Copying permitted for purchasing school only. This material is not copyright free.

1000	5
2000	4
3000	3
4000	2
5000	1

Abacus Ginn and Company 2001. Copying permitted for purchasing school only. This material is not copyright free.

6000	6
7000	8
8000	7
9000	9

Abacus Ginn and Company 2001. Copying permitted for purchasing school only. This material is not copyright free.

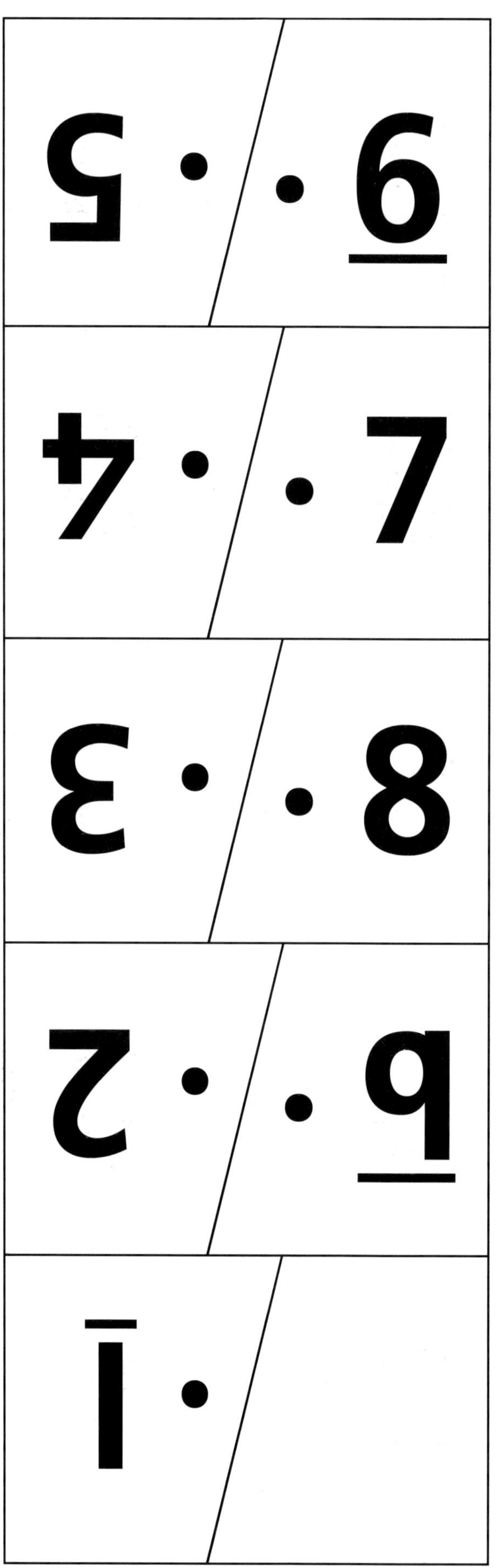

Abacus Ginn and Company 2001. Copying permitted for purchasing school only. This material is not copyright free.

·05	·09
·04	·07
·03	·08
·02	·09
·01	

Abacus Ginn and Company 2001. Copying permitted for purchasing school only. This material is not copyright free.

·005	·006
·004	·007
·003	·008
·002	·009
·001	

Abacus Ginn and Company 2001. Copying permitted for purchasing school only. This material is not copyright free.

⁻1	⁻2	⁻3
⁻4	⁻5	⁻<u>6</u>
⁻7	⁻8	⁻<u>9</u>
⁻10		

Abacus Ginn and Company 2001. Copying permitted for purchasing school only. This material is not copyright free.

0	1	2
3	4	5
6	7	8
9	10	

Abacus Ginn and Company 2001. Copying permitted for purchasing school only. This material is not copyright free.

11	12	13
14	15	16
17	18	19
20		

Abacus Ginn and Company 2001. Copying permitted for purchasing school only. This material is not copyright free.

21	22	23	24
25	26	27	28
29	30	31	32
33	34	35	36

Abacus Ginn and Company 2001. Copying permitted for purchasing school only. This material is not copyright free.

37	38	39	40
41	42	43	44
45	46	47	48
49	50	51	52

Abacus Ginn and Company 2001. Copying permitted for purchasing school only. This material is not copyright free.

53	54	55	56
57	58	59	<u>60</u>
<u>61</u>	62	63	64
65	<u>66</u>	67	<u>68</u>

Abacus Ginn and Company 2001. Copying permitted for purchasing school only. This material is not copyright free.

69	70	71	72
73	74	75	76
77	78	79	80
81	82	83	84

Abacus Ginn and Company 2001. Copying permitted for purchasing school only. This material is not copyright free.

85	86	87	88
89	90	91	92
93	94	95	96
97	98	99	100

Abacus Ginn and Company 2001. Copying permitted for purchasing school only. This material is not copyright free.

T (Tens)	U (Units)	• t (tenths)

Abacus Ginn and Company 2001. Copying permitted for purchasing school only. This material is not copyright free.

1	2	3	4	5	6	7	8	9	10
2	4	6	8	10	12	14	16	18	20
3	6	9	12	15	18	21	24	27	30
4	8	12	16	20	24	28	32	36	40
5	10	15	20	25	30	35	40	45	50
6	12	18	24	30	36	42	48	54	60
7	14	21	28	35	42	49	56	63	70
8	16	24	32	40	48	56	64	72	80
9	18	27	36	45	54	63	72	81	90
10	20	30	40	50	60	70	80	90	100

Abacus Ginn and Company 2001. Copying permitted for purchasing school only. This material is not copyright free.

1	2	3	4	5	6	7	8	9	10
11	12	13	14	15	16	17	18	19	20
21	22	23	24	25	26	27	28	29	30
31	32	33	34	35	36	37	38	39	40
41	42	43	44	45	46	47	48	49	50
51	52	53	54	55	56	57	58	59	60
61	62	63	64	65	66	67	68	69	70
71	72	73	74	75	76	77	78	79	80
81	82	83	84	85	86	87	88	89	90
91	92	93	94	95	96	97	98	99	100

Abacus Ginn and Company 2001. Copying permitted for purchasing school only. This material is not copyright free.

$\frac{2}{3}$	$\frac{3}{4}$	$\frac{3}{5}$
$\frac{4}{7}$	$\frac{3}{8}$	$\frac{5}{9}$
$\frac{7}{10}$	$\frac{2}{5}$	$\frac{5}{6}$
$\frac{7}{8}$	$\frac{1}{4}$	$\frac{1}{2}$
$\frac{1}{6}$	$\frac{1}{3}$	

Abacus Ginn and Company 2001. Copying permitted for purchasing school only. This material is not copyright free.

10%	20%	25%
30%	33%	40%
50%	60%	66%
70%	75%	80%
90%		

Abacus Ginn and Company 2001. Copying permitted for purchasing school only. This material is not copyright free.

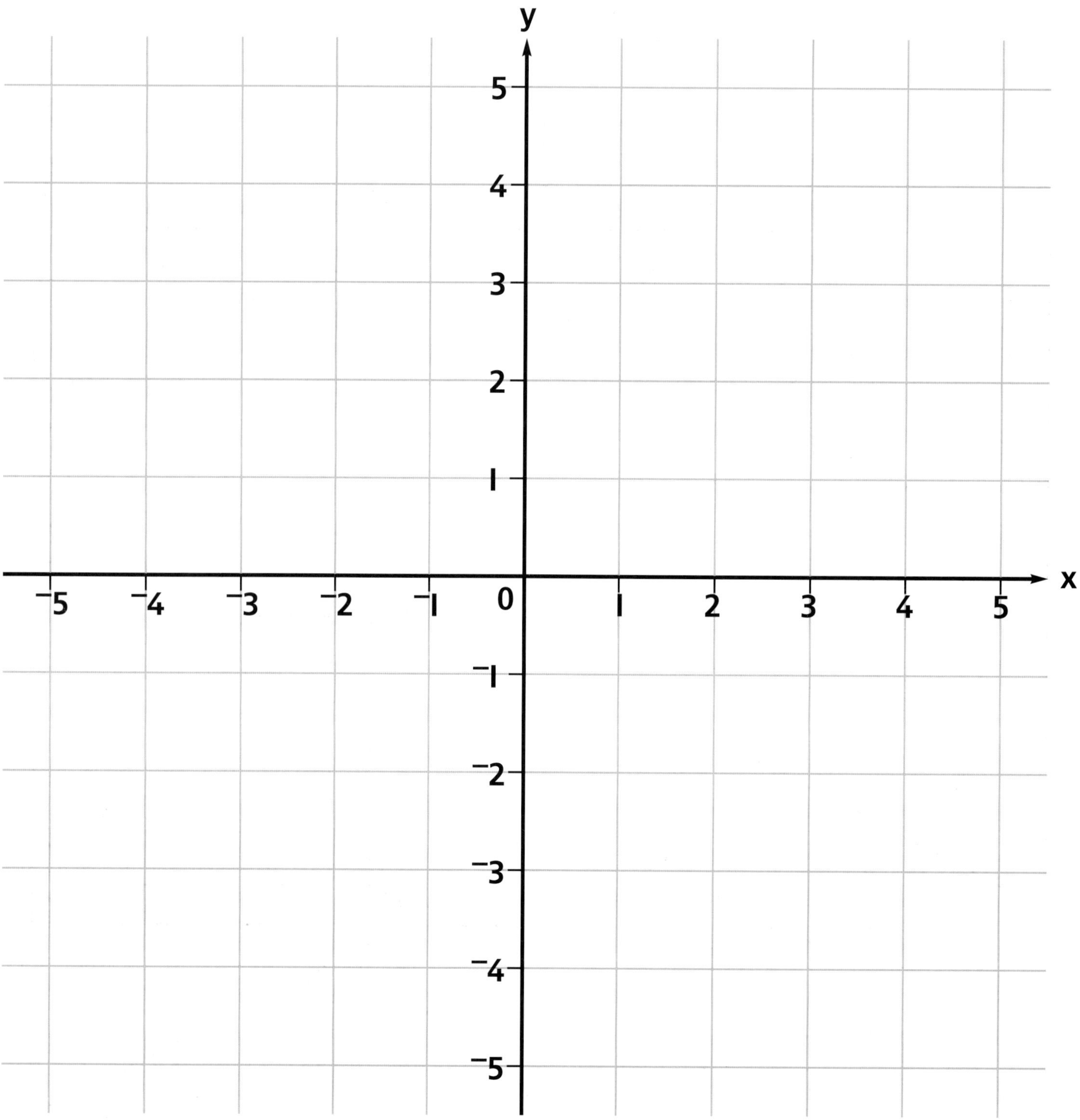

Abacus Ginn and Company 2001. Copying permitted for purchasing school only. This material is not copyright free.

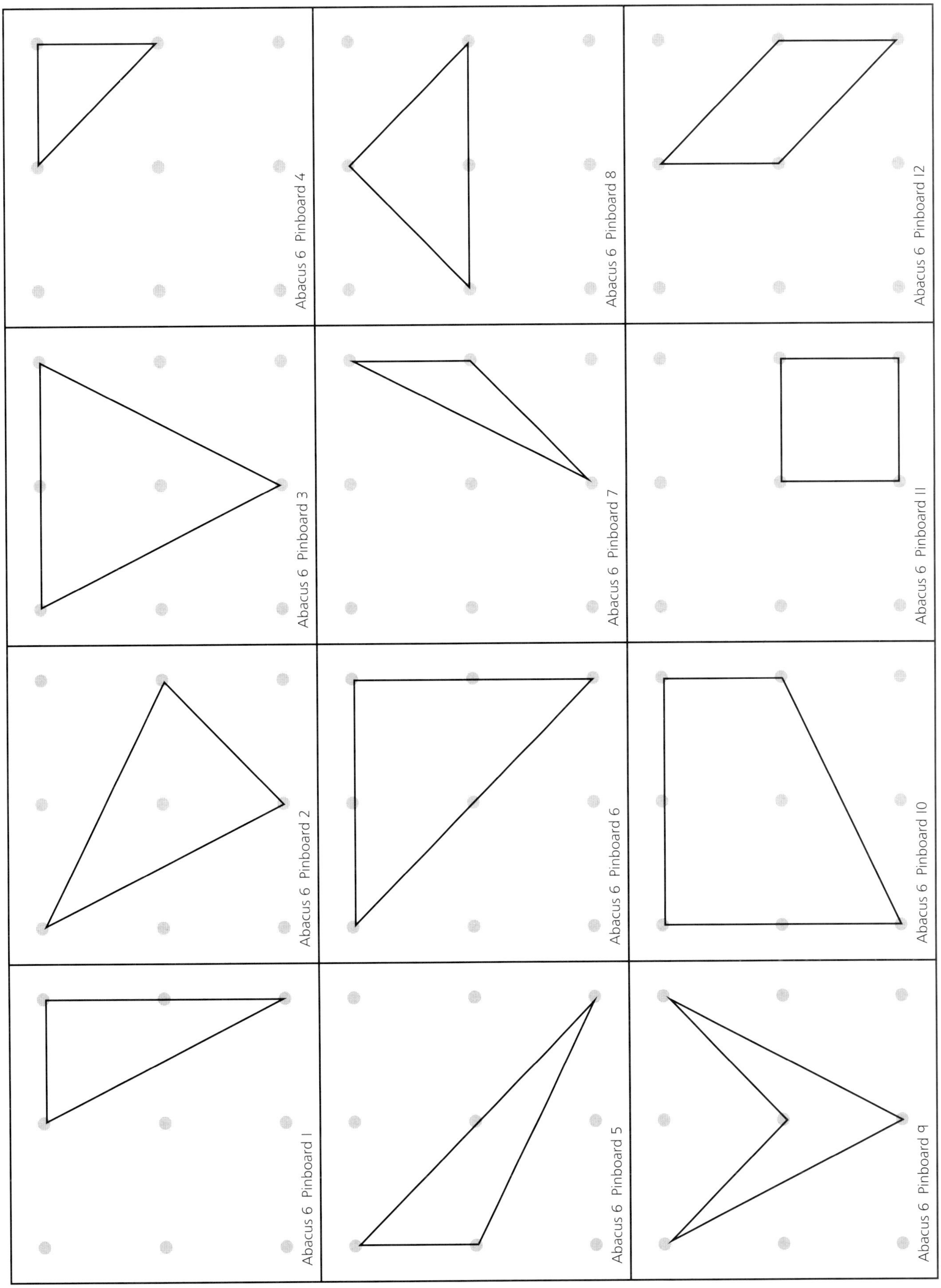

Abacus 6 Pinboard 4	Abacus 6 Pinboard 8	Abacus 6 Pinboard 12
Abacus 6 Pinboard 3	Abacus 6 Pinboard 7	Abacus 6 Pinboard 11
Abacus 6 Pinboard 2	Abacus 6 Pinboard 6	Abacus 6 Pinboard 10
Abacus 6 Pinboard 1	Abacus 6 Pinboard 5	Abacus 6 Pinboard 9

Abacus Ginn and Company 2001. Copying permitted for purchasing school only. This material is not copyright free.

Pinboard shape cards (sheet 2)

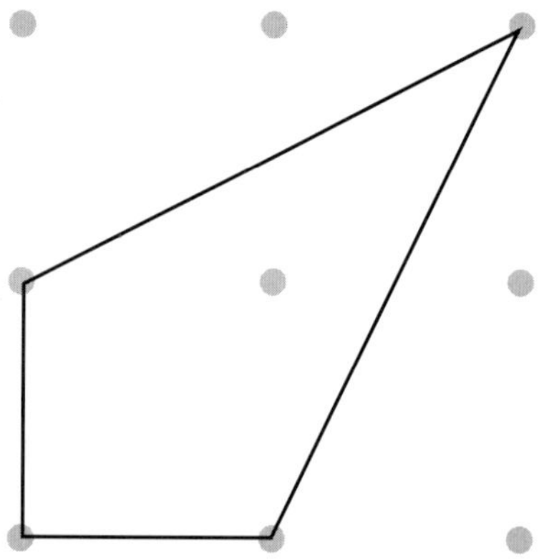

Abacus 6 Pinboard 16	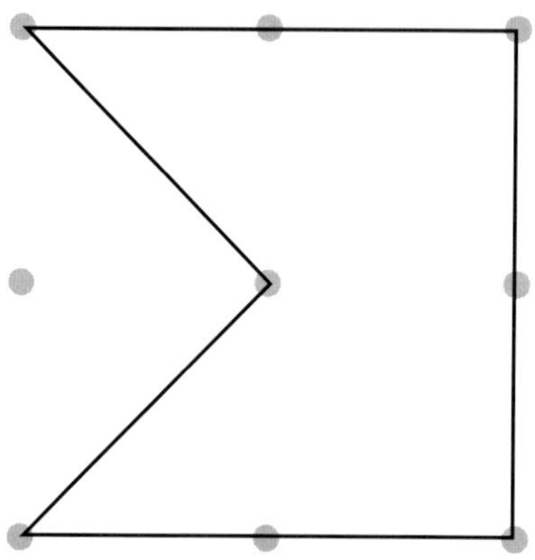 Abacus 6 Pinboard 20	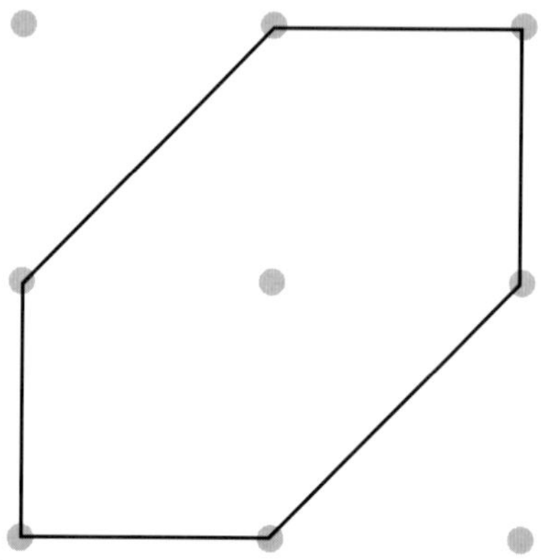 Abacus 6 Pinboard 24
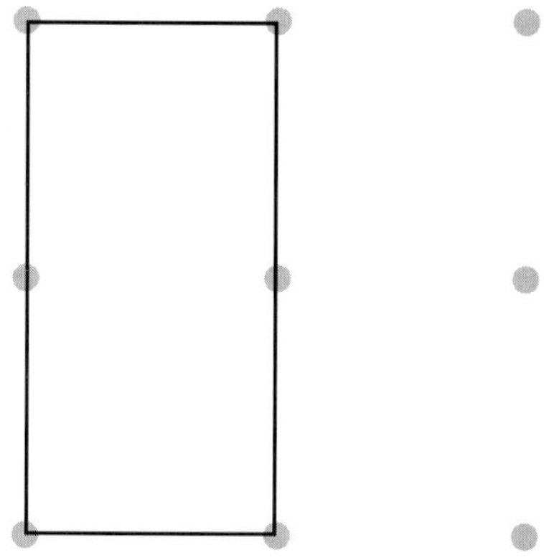 Abacus 6 Pinboard 15	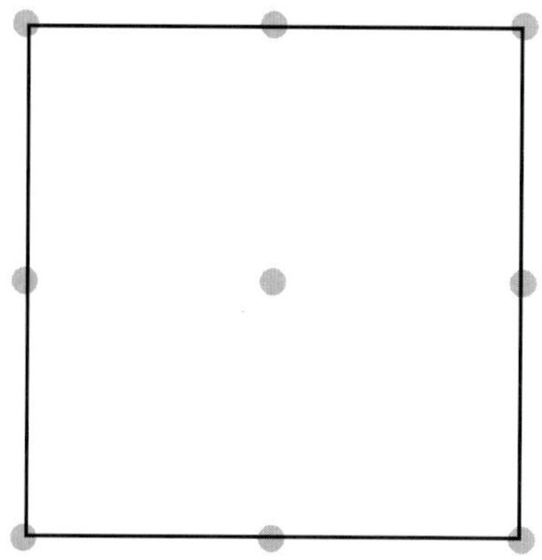 Abacus 6 Pinboard 19	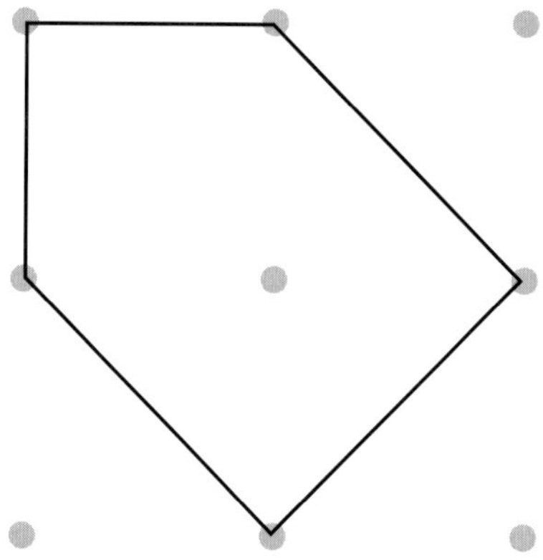 Abacus 6 Pinboard 23
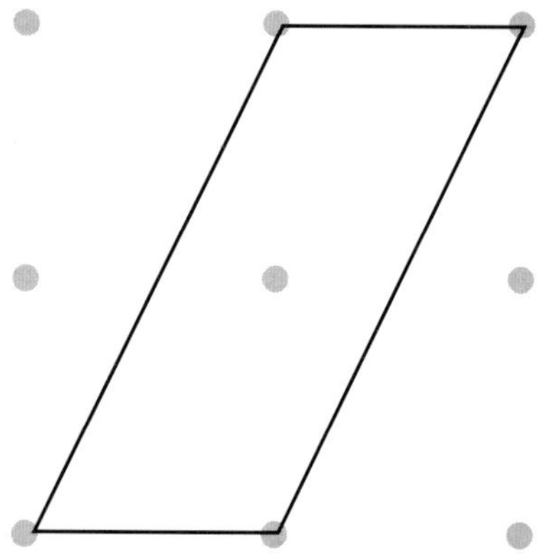 Abacus 6 Pinboard 14	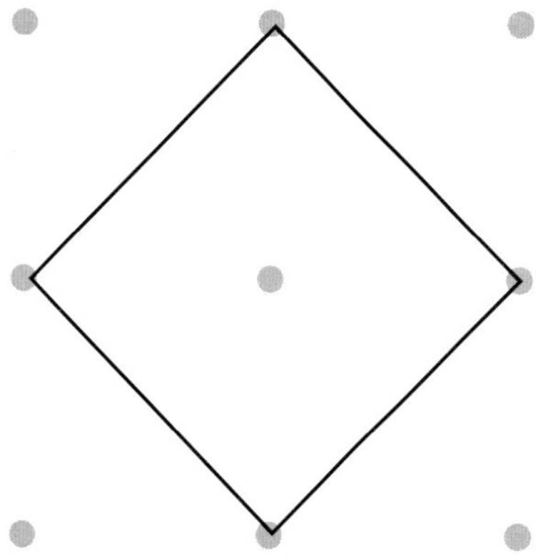 Abacus 6 Pinboard 18	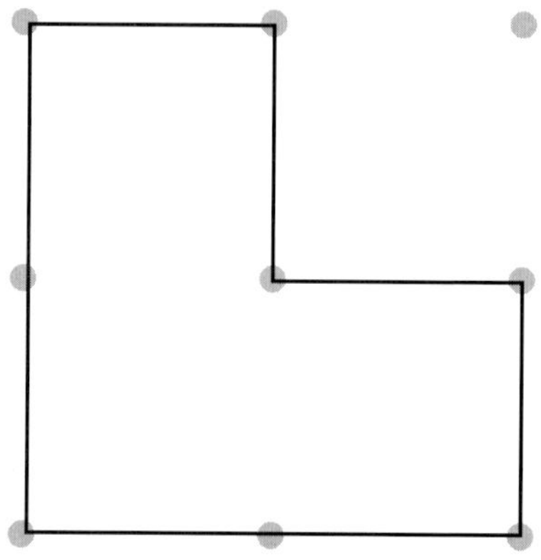 Abacus 6 Pinboard 22
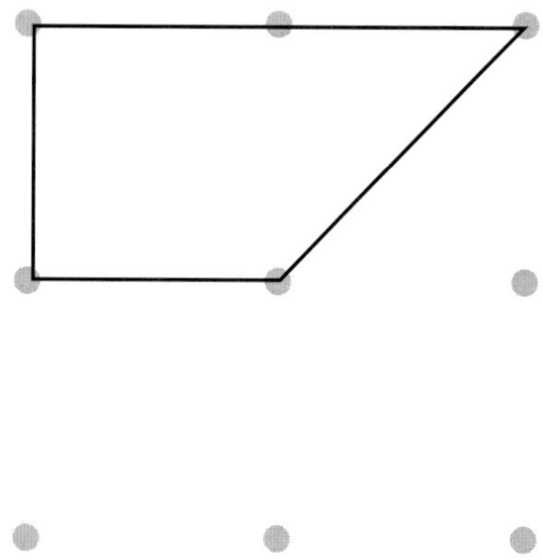 Abacus 6 Pinboard 13	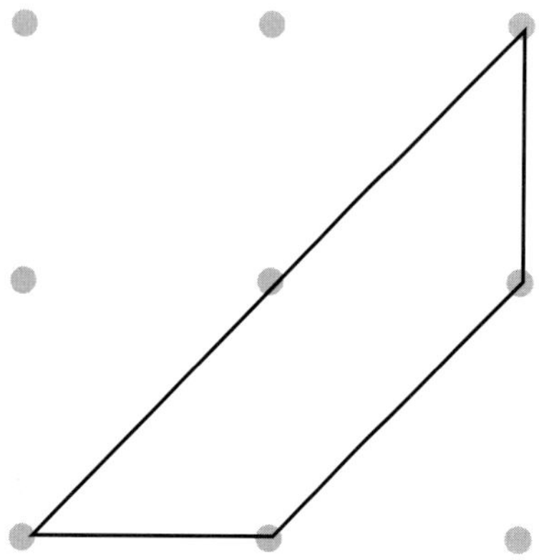 Abacus 6 Pinboard 17	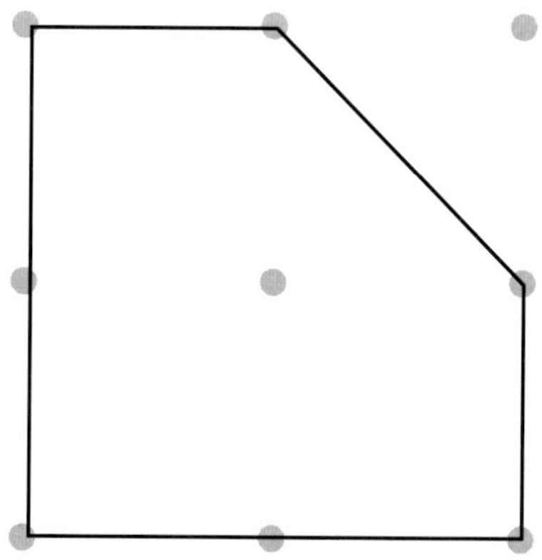 Abacus 6 Pinboard 21

Abacus Ginn and Company 2001. Copying permitted for purchasing school only. This material is not copyright free.

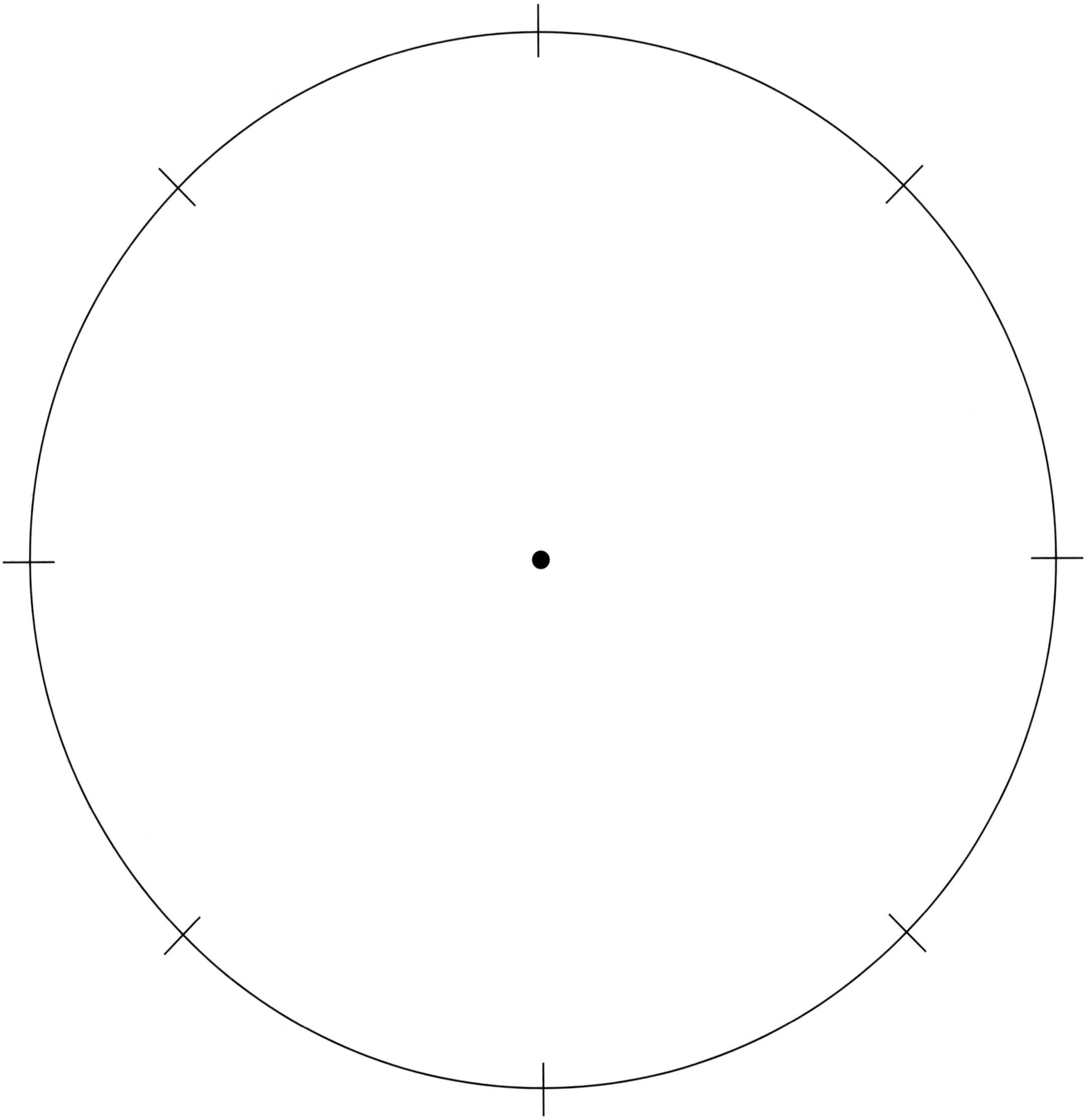

Abacus Ginn and Company 2001. Copying permitted for purchasing school only. This material is not copyright free.

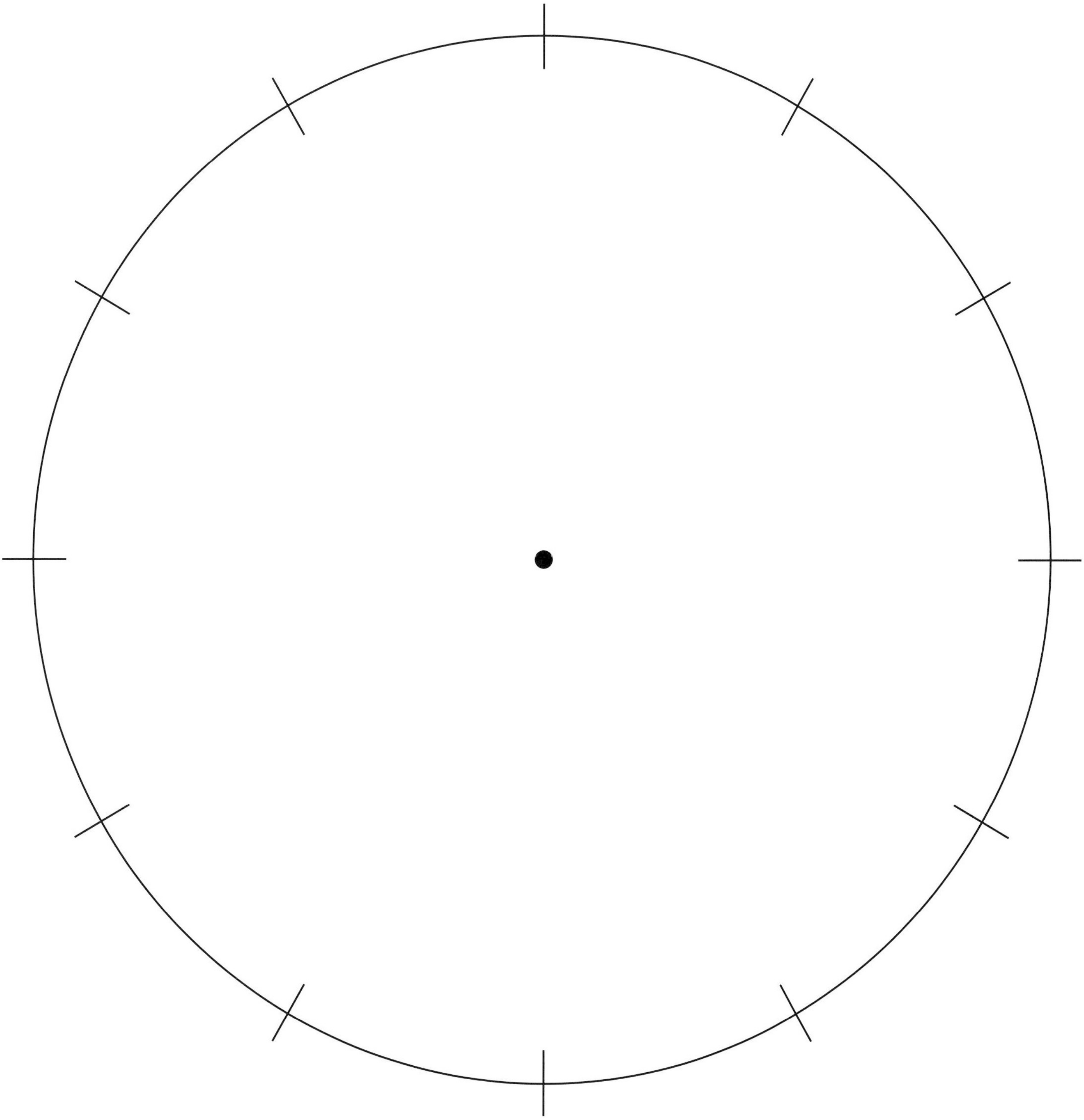

Abacus Ginn and Company 2000. Copying permitted for purchasing school only. This material is not copyright free.

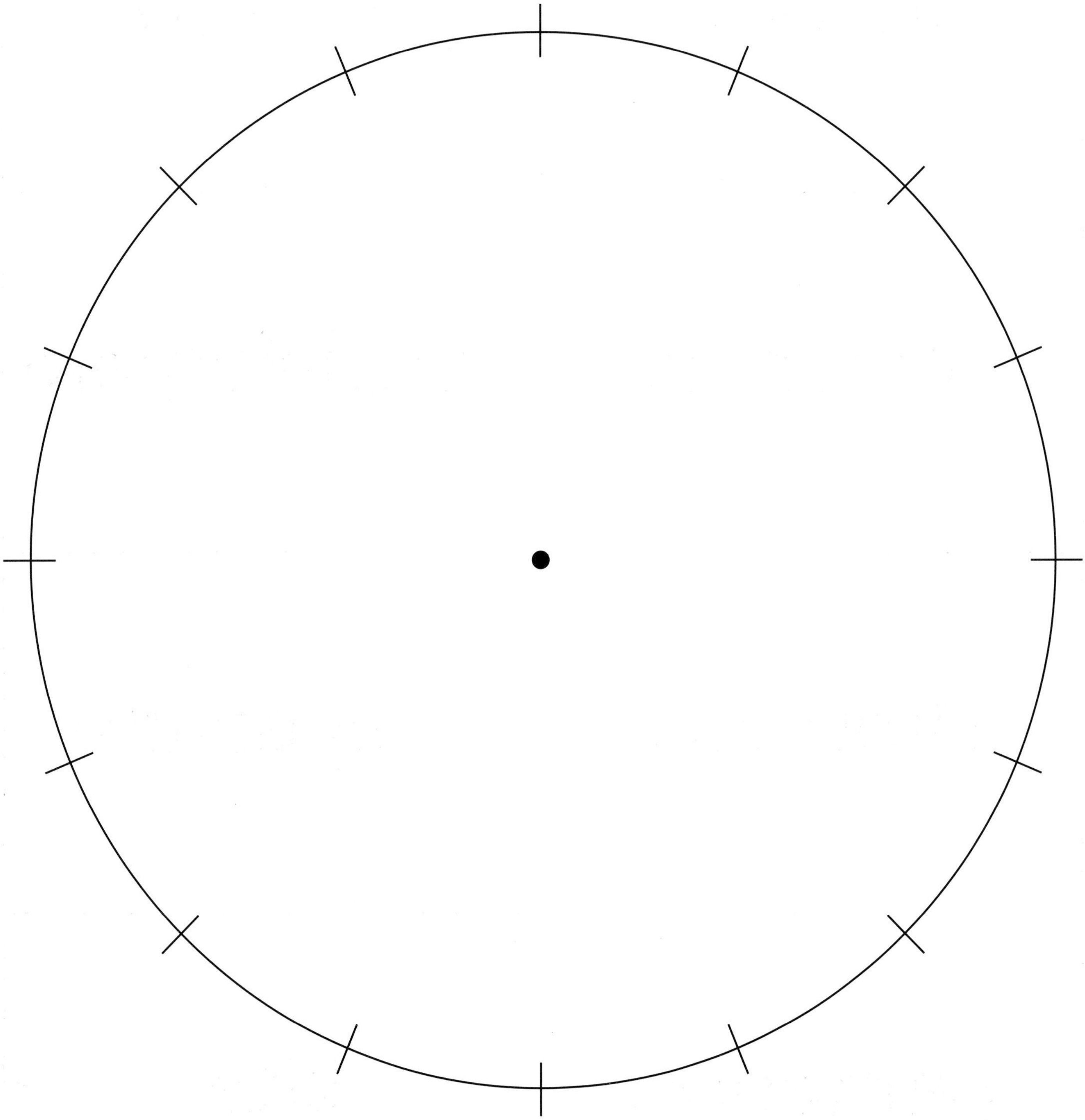

Abacus Ginn and Company 2000. Copying permitted for purchasing school only. This material is not copyright free.

square	**triangle**
rectangle	**parallelogram**
rhombus	**trapezium**
pentagon	**kite**

Abacus Ginn and Company 2001. Copying permitted for purchasing school only. This material is not copyright free.

hexagon	**quadrilateral**
heptagon	**octagon**
nonagon	**decagon**

Abacus Ginn and Company 2001. Copying permitted for purchasing school only. This material is not copyright free.

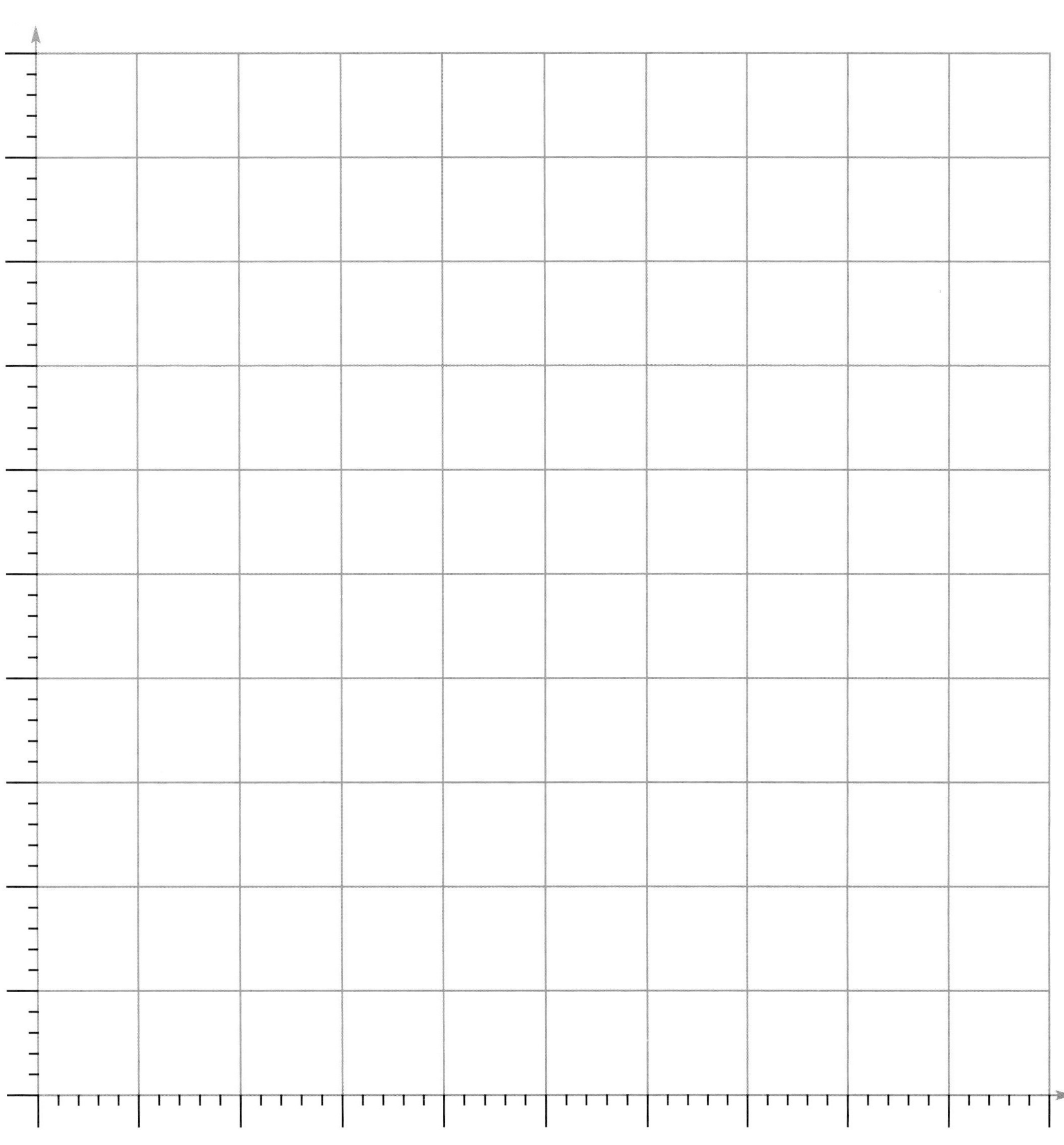

Abacus Ginn and Company 2001. Copying permitted for purchasing school only. This material is not copyright free.